AI at the Wheel

The Revolution of Autonomous Driving

by
Derek Lawson

AI at the Wheel

The Revolution of Autonomous Driving

Contents

Introduction

Autonomous driving, once a figment of science fiction, is now a rapidly evolving reality. As artificial intelligence (AI) increasingly integrates into various aspects of our lives, its impact on the transportation industry is both profound and multifaceted. This book aims to unravel the complexities of this transformation, providing readers with a comprehensive understanding of how AI is revolutionizing our roads and reshaping societal norms, economic landscapes, and ethical frameworks.

The potential of autonomous vehicles (AVs) extends beyond the convenience of a hands-free driving experience. It promises to redefine urban planning, reduce traffic accidents, lower emissions, and even alter our relationship with mobility. As such, the implications are vast, spanning from technological advancements to societal shifts. Yet, with this promise comes a range of challenges that need to be addressed—from technology and infrastructure to ethics and legalities. We will delve into these areas through an array of perspectives and expert insights.

In the pursuit of knowledge, it's essential to understand the foundational elements that underpin autonomous driving. This journey begins with a clear definition of what constitutes an AV and a review of the pioneering efforts that brought us to where we are today. Early innovations in self-driving technology laid the groundwork for the sophisticated systems now nearing mainstream acceptance.

At the heart of autonomous vehicles are advanced sensors, hardware, and complex algorithms powered by AI. These components work together seamlessly to interpret the environment and make real-time decisions, which are critical for safe and efficient driving. An exploration of these technologies reveals not just the engineering marvels behind them but also the challenges they overcome in the quest for reliability and safety.

The journey from rudimentary autonomous systems to the advanced models we see today is marked by significant historical milestones and key contributions from influential researchers and companies. Understanding this evolution is crucial for appreciating the current state of AI in transportation and envisioning its future trajectory.

One cannot ignore the sociocultural impact of AVs. As vehicles gain the ability to drive themselves, they change the fabric of our cities and the way we interact with urban spaces. The public's perception and acceptance of these changes are pivotal to their success. This book will explore how societal attitudes shape the adoption of autonomous vehicles and, conversely, how AVs influence social norms and lifestyles.

Economically, the introduction of autonomous driving is poised to bring about significant shifts. The job market, particularly for drivers, will see substantial changes, while new opportunities in tech development and related fields will arise. Additionally, market growth and investment trends will be examined to provide a holistic view of the economic landscape shaped by AVs.

The ethical considerations of autonomous driving present some of the most nuanced and critical debates in this field. Moral dilemmas, such as decision-making in unavoidable accident scenarios, raise questions about the programming of AI and the accountability of developers. Regulatory and legal challenges also arise as lawmakers grapple with the implications of autonomous technology, striving to create frameworks that ensure safety and fairness.

As with any technology reliant on vast amounts of data, big data plays a crucial role in the functionality and improvement of autonomous vehicles. From data collection methods to real-time processing and analytics, the importance of managing and interpreting data cannot be overstated. This book will discuss how big data drives the precision and responsiveness of AVs.

Safety and reliability are paramount in the deployment of self-driving cars. Collision avoidance systems, alongside exhaustive validation and testing protocols, ensure that these vehicles can operate under various conditions without compromising passenger safety. The book will cover the methods and technologies employed to achieve these standards.

Vehicle-to-Everything (V2X) communication stands as another pillar of autonomous driving. The interconnectedness of vehicles with infrastructure, pedestrians, and other elements of the transport ecosystem is crucial for seamless operation. We'll discuss the infrastructure requirements, performance metrics, and standards required to make V2X communication effective.

Environmental considerations are also at the forefront. Autonomous vehicles have the potential to significantly reduce emissions and promote the use of sustainable materials and technologies. This book will explore how these innovations contribute to a greener future.

Public transportation systems are not left untouched by the wave of automation. Autonomous buses and shuttles represent a new frontier in public mobility, promising to integrate smoothly into existing systems. An understanding of how these vehicles will enhance public transit is essential for grasping the broader impact of autonomous technology.

The rise of autonomous vehicles also brings new dynamics to ride-sharing services. On-demand services and novel business models are emerging, transforming the economics of personal and shared mobility. This book will delve into how these models are evolving and the implications for both consumers and companies.

Government and policy-making play an instrumental role in the development and deployment of AVs. Regulation, licensing, and international collaboration on standards are critical components that facilitate the safe and widespread adoption of autonomous technologies. The interplay between government actions and technological advancements will be examined.

Cybersecurity remains a vital concern in the realm of autonomous vehicles. Protecting AVs against potential threats and ensuring secure communication protocols are imperative for maintaining trust and reliability. This book will cover strategies and technologies designed to safeguard these systems.

Fleet management is another area where AI is making significant strides. Optimization techniques and predictive maintenance are just a few examples of how AI can improve efficiency and reduce costs for fleet operators. The book will explore these benefits and their practical applications.

The deployment of autonomous vehicles presents unique challenges in rural versus urban settings. The infrastructure differences and specific solutions required for each environment will be discussed, highlighting how AVs can cater to diverse geographical needs.

Personal car ownership is expected to undergo a transformation with the emergence of self-driving technology. Changes in consumer behavior, future sales trends, and ownership models will be examined to provide a glimpse into what lies ahead.

Legal frameworks and liabilities are continually evolving in response to advancements in autonomous driving. The responsibility in accident scenarios and the development of new legal precedents are critical issues that this book will address, offering insight into the complexities of law in the age of AI.

Insurance models and policies will also need to adapt to the unique features of autonomous vehicles. Risk assessment techniques will evolve, demanding a thorough analysis of how insurance will keep pace with technological advancements.

Navigation and mapping are fundamental to the operation of AVs. With real-time updates and advanced routing algorithms, these vehicles can navigate efficiently and safely. This book will cover the technologies driving these capabilities.

Human-machine interaction is another critical aspect of autonomous driving. User interface design and improving trust and comfort between humans and machines are key areas of focus, ensuring that users feel secure and confident in using AVs.

Autonomous commercial vehicles, including trucks and delivery systems, represent a significant sector where AI can bring economic benefits and logistical advantages. The operational and economic challenges and benefits will be discussed in depth.

The role of startups and innovators cannot be overstated. Disruptive technologies and the investment landscape are driving significant advancements in the field of autonomous vehicles. This book will highlight the contributions and challenges faced by these entities.

Long-term predictions and trends offer a glimpse into the future of autonomous

Chapter 1:
The Dawn of Autonomous Driving

The rise of autonomous driving marks a transformative moment in the history of transportation, reshaping our understanding of mobility and convenience. Emerging from decades of research and innovation, autonomous vehicles (AVs) are no longer a speculative dream but a burgeoning reality. With sensors and advanced AI algorithms at their core, these self-driving marvels promise to reduce accidents, ease traffic congestion, and offer greater accessibility. Pioneers in the field have laid the groundwork, but as we transition from prototype to mainstream adoption, the implications stretch far beyond technology. This chapter explores how autonomy in vehicles is set to revolutionize our cities, impact societal norms, and redefine what we consider possible in personal and public transportation. The dawn of autonomous driving is upon us, ushering in an era of unprecedented change.

Defining Autonomous Vehicles

At its core, the term "autonomous vehicle" refers to a car capable of navigating and operating without human intervention. This advancement pivots from traditional automotive paradigms where human drivers made real-time decisions based on environmental stimuli and road conditions. Instead, these vehicles rely on an array of sophisticated technologies, including sensors, machine learning, and artificial intelligence, to perceive their surroundings and execute driving tasks safely.

Self-driving cars aren't a monolithic concept but rather exist on a spectrum of automation. The Society of Automotive Engineers (SAE) defines six levels of automation, ranging from Level 0, where the human driver is fully in control, to Level 5, where the car is completely autonomous in all conditions. This framework helps us understand the gradual journey from semi-autonomous features like adaptive cruise control and lane-keeping assistance to full autonomy, where human presence is optional.

Level 1 automation includes systems that provide minor driving assistance. A common example is adaptive cruise control, which maintains a set speed and adjusts to the speed of the vehicle in front. At Level 2, drivers encounter partial autonomy; vehicles can manage both steering and speed under certain conditions, although the driver must remain alert and ready to take over.

It's at Levels 3 and 4 where things start to get truly interesting. Level 3 cars can make decisions autonomously in specific environments, like highway driving, but may still require human intervention during complex situations. Level 4 vehicles can handle all driving tasks within designated areas and conditions, known as Operational Design Domains (ODDs). However, Level 4 cars may still need human control in scenarios outside their ODD. Finally, Level 5 vehicles represent the pinnacle of autonomy, capable of handling all driving tasks in all conditions without any human input.

The essence of autonomous driving lies in complex software and hardware systems working in harmony. Central to this is the car's "perception stack," an intricate network of sensors like LIDAR, radar, and cameras, collaborating to interpret surrounding environments. These sensors generate massive amounts of data, which AI algorithms process in real-time to make split-second decisions—a feat previously reserved for human cognition.

LIDAR, short for Light Detection and Ranging, uses laser pulses to create detailed 3D maps of the vehicle's surroundings. Radar complements this by detecting objects' distance and speed, crucial for collision avoidance. Cameras provide high-resolution images for recognizing lane markings, traffic signals, and pedestrians. Combined, these technologies offer a comprehensive situational awareness that's critical for safe autonomous operation.

Equally important is the vehicle's decision-making software. Machine learning algorithms, particularly those involving deep learning, enable the car to recognize patterns and predict future events. For instance, these algorithms can identify pedestrian movement trends to anticipate crossing behavior or recognize changes in traffic light patterns. Over time, as these systems "learn" from vast datasets, their predictive accuracy and reliability continually improve.

Data is the lifeblood of autonomous vehicles. Without continuous streams of data, these cars wouldn't be able to adapt to new environments or improve over time. This data comes from multiple sources: from sensor inputs during live operations to extensive simulations replicating countless driving scenarios. It's then fed into neural networks that refine the vehicle's driving capabilities, learning nuances like how to handle complex intersections or navigate through dense urban areas.

Autonomous vehicles also embody a paradigm shift in vehicle connectivity. They're often integrated with Vehicle-to-Everything (V2X) communication systems, which enable them to interact with other vehicles, infrastructure, and even pedestrians. V2X communication allows for more coordinated and efficient traffic flow, enhancing overall safety and mobility.

AI at the Wheel

However, defining autonomous vehicles extends beyond just technical prowess. It includes understanding their potential impact on society and the economy. These vehicles promise to revolutionize transport by reducing accidents caused by human error, alleviating traffic congestion, and providing greater mobility for individuals unable to drive, such as the elderly or disabled. Furthermore, they hold the key to optimizing logistic operations, reducing energy consumption, and cutting greenhouse gas emissions through more efficient driving patterns.

Despite these promises, several hurdles remain. Ensuring the reliability and safety of these vehicles is paramount. Autonomous vehicles must be tested rigorously across a multitude of environments and conditions. Moreover, there's the challenge of public acceptance, a critical factor for widespread adoption. Trust must be earned—people need to see these vehicles operating safely and reliably in real-world conditions.

Regulatory landscapes will also play a crucial role in defining the future of autonomous vehicles. Policymakers must strike a careful balance between fostering innovation and ensuring safety. Regulations governing the use of autonomous vehicles will need to evolve continually to address new challenges and opportunities as the technology matures. This regulatory evolution will likely differ across regions, reflecting varying levels of technological readiness and societal acceptance.

Lastly, ethical considerations can't be ignored. The deployment of autonomous vehicles brings forth complex moral dilemmas. For instance, how should an autonomous car prioritize lives in a potential accident involving multiple parties? Resolving such ethical quandaries demands a collaborative effort involving technologists, ethicists, and policymakers.

In conclusion, defining autonomous vehicles involves a multifaceted understanding of technological innovations, regulatory frameworks, sociocultural impacts, and ethical dimensions. These vehicles represent not just a leap in automotive technology but a broader shift towards a more connected, efficient, and inclusive transportation ecosystem. As we continue to explore and develop this transformative technology, the focus will be on creating a future where autonomous vehicles enhance our lives and redefine mobility as we know it.

Early Innovators in Self-Driving Technology

The advent of autonomous driving technology has not only captivated imaginations but also set the stage for transformative shifts in transportation. In the early days, when the concept of self-driving cars was still in its infancy, several innovators and companies played pivotal roles in laying the groundwork. These early pioneers made substantial contributions that have shaped the trajectory of autonomous vehicle development, pushing the boundaries of what was thought possible at the time.

One of the most recognized names in the realm of self-driving technology is Google, now known under its parent company, Alphabet. Google launched its self-driving car project, Waymo, in 2009. The project's goal was straightforward yet ambitious: to drive fully autonomously on numerous types of roads and conditions without any human intervention. Google's research added invaluable data to the field, accelerating advancements in AI, sensor technology, and real-time data processing. Their fleet of modified Toyota Priuses and later, the custom-built "Firefly," navigated city streets and highways, gathering millions of miles of driving data essential for refining algorithms and improving safety features.

Tesla, another notable pioneer, took a different approach than Google's comprehensive autonomous system. Instead of aiming for full autonomy from the get-go, Tesla focused on an incremental, Autopilot system designed to assist drivers. Tesla's CEO, Elon Musk, envisioned a future where over-the-air updates could continuously improve the car's capabilities. Tesla's use of a vast sensor suite, including cameras, radar, and ultrasonic sensors, positioned it to be at the forefront of gradually achieving higher levels of autonomy, guided by real-world data from millions of cars on the road.

While Google and Tesla might be household names, several other early initiatives were just as crucial. Stanford University's Stanley, a robotic vehicle, won the 2005 DARPA Grand Challenge, an event that undoubtedly boosted public and industry confidence in the feasibility of autonomous driving. Stanley's success underscored the importance of robust AI systems capable of making split-second decisions and adapting to unpredictable environments. This win marked a significant milestone and brought academic rigor and credibility to the burgeoning field.

Ford Motor Company also dipped its toes into the self-driving arena quite early on. In partnership with the University of Michigan, Ford tested its autonomous prototypes on the Mcity facility's roads, specially designed for testing connected and automated vehicles. Their methodical approach involved comprehensive mapping, radar and LiDAR technologies, and a keen focus on vehicle-to-vehicle communication. Ford's structured focus on both technological development and regulatory aspects has made a significant contribution to balancing innovation with safety standards.

Nissan, too, embarked on its journey with ambitious objectives. They installed a group of engineers in Silicon Valley, a hub for tech innovation, aiming to create a car that would navigate highways independently by 2020. Although they missed this self-imposed deadline, their intensive research and development efforts resulted in advanced driver assistance systems that have filtered down into consumer cars, enhancing safety and driver convenience.

On a different note, Uber, although better known for its ride-sharing platform, made significant strides in autonomous technology through its Advanced Technologies Group. They tackled the issue from a service-oriented perspective, emphasizing how self-driving cars could disrupt and fundamentally change urban transportation networks. Despite facing setbacks, including fatal accidents that raised ethical and safety concerns, Uber's data and advancements have continued to influence discussions on the safe deployment of autonomous vehicles.

Pioneers are often driven by innovation, but regulations and societal impact can equally shape their paths. Mercedes-Benz demonstrated its vision through the Mercedes F 015 Luxury in Motion, a concept car unveiled in 2015. This vehicle was designed from the ground up to challenge both technology and conventional perception, incorporating futuristic interiors to envision how people might interact within an autonomous vehicle. Mercedes-Benz's holistic approach extended beyond just driving; it considered the broader sociocultural shifts that such technology would introduce.

Similar innovative strides were made by Volvo, who partnered with Uber for a joint initiative in developing self-driving capabilities. Their focus leaned heavily towards incorporating redundant systems—essentially backup mechanisms that could take over if primary systems failed. This focus highlighted the critical importance of reliability and system integrity in building public trust and ensuring safety in real-world applications.

Lastly, we cannot overlook the contributions from academia and smaller startups that were equally transformative. Institutions like Carnegie Mellon University (CMU) have long been involved in robotics and AI, frequently collaborating with both government and private enterprises. Their research and resultant technologies have often found their way into commercial and military applications, showcasing the tangible impact of academic endeavors.

Startups like Cruise Automation, later acquired by General Motors, injected fresh ideas and agility into the industry. Cruise's focused approach on urban environments enabled them to develop systems optimized for the complexities of city driving. This acquisition illustrated the trend of legacy automotive companies absorbing innovative startups to fuse traditional expertise with cutting-edge technology.

In retrospect, these early innovators didn't merely contribute to technological advances but also helped foster a competitive environment that accelerated growth. They facilitated invaluable discussions on ethics, safety, and regulatory frameworks, setting the stage for current and future advancements in autonomous vehicle technology. As we continue to explore the uncharted territories of self-driving cars, recognizing the efforts of these early pioneers provides a comprehensive view of how we arrived at this pivotal juncture in transportation history.

From sweeping test tracks to intricate urban landscapes, these trailblazers laid the groundwork. Their collective efforts have undeniably charted a path towards a future where autonomous vehicles could become a commonplace reality, profoundly affecting how we perceive and interact with transportation systems globally.

Chapter 2:
The Technology Behind
Self-Driving Cars

The technology propelling self-driving cars is a seamless integration of advanced sensors, high-speed processors, and intelligent algorithms. At the heart of autonomous driving lie an array of sensors, such as LiDAR, cameras, and radar, which continuously gather data about the vehicle's surroundings. This raw data is then processed using sophisticated machine learning and AI algorithms, enabling the car to recognize objects, make decisions, and navigate safely. These algorithms learn from vast amounts of driving data, refining their ability to handle complex scenarios on the road. Coupled with robust hardware, self-driving cars leverage real-time data analysis and predictive modeling to create an intricate map of their environment, allowing for precise and adaptive driving. This confluence of technology ensures that autonomous vehicles can operate with a level of safety and efficiency that was once the realm of science fiction, bringing us closer to a future where transportation is both smarter and more reliable.

Sensors and Hardware

The foundation of any self-driving car is its array of sensors and hardware components. These essential pieces of technology are akin to the human senses, enabling the vehicle to perceive and interpret its surroundings. The combination of various sensors ensures a comprehensive understanding of the environment, facilitating safe and efficient navigation.

Lidar (Light Detection and Ranging) systems are among the most crucial sensors in autonomous vehicles. Lidar uses laser light to create high-resolution, three-dimensional maps of the car's surroundings. By emitting rapid pulses of light and measuring the time it takes for each pulse to reflect back from an object, Lidar can construct a detailed spatial representation. This real-time mapping technology excels in detecting obstacles, pedestrians, and other vehicles, even in low-light conditions.

However, Lidar isn't the only sensor at work. Cameras play a vital role by providing visual data, much like human eyes. These high-resolution cameras capture images of the road, traffic signs, and other critical elements. They work in tandem with advanced image recognition algorithms to interpret the visual data and inform the car's decision-making processes.

Radar sensors complement the capabilities of Lidar and cameras by accurately measuring the speed and distance of objects. Unlike Lidar, radar is less affected by adverse weather conditions such as fog, rain, or snow. This robustness makes radar an invaluable tool for maintaining the car's awareness of its surroundings in various environments.

Ultrasonic sensors are another important component, primarily used for short-range detection. These sensors emit ultrasonic waves and measure their reflection to detect nearby objects. Typically positioned around the car's perimeter, ultrasonic sensors are particularly useful for low-speed maneuvers, such as parking.

While each type of sensor offers unique advantages, the true strength of autonomous systems lies in sensor fusion. Sensor fusion refers to the integration of data from multiple sensors to create a more accurate and reliable picture of the environment. This process enhances the vehicle's ability to make informed decisions, ensuring a higher degree of safety and precision.

To support these sensors, self-driving cars are equipped with powerful computing hardware. Central to this setup is the onboard computer, often referred to as an Electronic Control Unit (ECU). The ECU's role is to process vast amounts of data from the sensors in real-time. This data processing involves complex algorithms that interpret sensory inputs, plan routes, and execute driving maneuvers.

These computers are equipped with specialized chips optimized for AI and machine learning tasks, such as Graphics Processing Units (GPUs) and Tensor Processing Units (TPUs). These chips are designed to handle the parallel processing required for tasks like image recognition and neural network computations. With their impressive computational capabilities, they enable the rapid decision-making critical for autonomous driving.

Connectivity hardware is another essential element, allowing self-driving cars to communicate with external systems and infrastructure. Vehicle-to-Everything (V2X) communication technology enables the exchange of information between the car and other vehicles, traffic signals, and road infrastructure. This connectivity enhances situational awareness and can lead to more coordinated and efficient traffic flows.

Moreover, redundant systems and fail-safes are integral to the hardware stack of autonomous vehicles. Redundancy ensures that if one component fails, others can take over to maintain safe operation. For example, if a primary sensor malfunctions, backup sensors can provide the necessary data. This redundancy is crucial for minimizing the risk of failures that could compromise safety.

In addition to the sensors themselves, the physical placement and integration of these components are meticulously designed. Lidar units, for instance, are often mounted on the roof to provide a 360-degree view. Cameras are strategically positioned around the vehicle to cover blind spots and ensure comprehensive visual coverage.

Furthermore, the durability and reliability of sensors and hardware are of paramount importance. These components must withstand diverse weather conditions, vibrations, and other environmental stresses. Rigorous testing and validation protocols are employed to ensure the robustness of these systems, aiming for high performance regardless of external factors.

Recent advancements in hardware miniaturization and cost reduction are making sensor technologies more accessible. Lidar, which was once prohibitively expensive, is becoming more affordable due to innovations in manufacturing and increasing demand. This trend is expected to accelerate the adoption of autonomous driving technology across different vehicle segments, from luxury cars to mass-market models.

The collaboration between automotive manufacturers and technology companies is also driving progress in sensor and hardware development. Partnerships enable the fusion of expertise from different domains, resulting in more sophisticated and integrated solutions. Companies are continuously exploring ways to improve sensor accuracy, reduce computational latency, and enhance overall system reliability.

In conclusion, the sensors and hardware components of self-driving cars form the backbone of their operational capabilities. By combining Lidar, cameras, radar, and ultrasonic sensors with powerful computing hardware and connectivity technologies, these vehicles achieve a high level of environmental awareness and decision-making acumen. As technology continues to evolve, advancements in sensors and hardware will play a pivotal role in shaping the future of autonomous transportation, making it safer, more efficient, and more accessible.

Machine Learning and AI Algorithms

Self-driving cars represent one of the most complex and impactful applications of machine learning and artificial intelligence (AI) algorithms in modern technology. At the core, these vehicles rely heavily on a variety of sophisticated techniques to interpret the world around them, make decisions, and execute actions without human intervention. To achieve such autonomy, self-driving cars must process vast amounts of data in real-time, detect and classify objects, predict future states, and plan safe paths.

Deep learning, a subset of machine learning, plays a pivotal role in teaching autonomous vehicles to understand their environment. Neural networks, designed to mimic the human brain's structure, are trained on large datasets comprising diverse driving conditions and scenarios. These networks are particularly effective in recognizing patterns and making predictions based on those patterns. For instance, convolutional neural networks (CNNs) are commonly used for image recognition tasks like identifying lanes, traffic signs, and pedestrians.

One critical aspect is computer vision, which enables the car to "see" its surroundings. Cameras mounted on the vehicle capture real-time images, which are then processed using computer vision algorithms. These algorithms convert visual data into numerical formats that the AI can interpret. Object detection frameworks such as YOLO (You Only Look Once) and Faster R-CNN (Region-Based Convolutional Neural Networks) are essential in identifying and classifying objects in the vehicle's path. They must also distinguish between moving and stationary objects, an important factor in decision-making.

Another cornerstone of self-driving technology is sensor fusion, which amalgamates data from various sensors for a holistic understanding of the environment. While cameras provide visual data, other sensors like LiDAR (Light Detection and Ranging) and RADAR (Radio Detection and Ranging) supply additional layers of information. LiDAR uses laser beams to create high-resolution 3D maps, essential for precise measurements of distances and object contours. By combining these multiple data sources, machine learning algorithms can offer a more accurate perception of the surroundings, compensating for the limitations of individual sensors.

Reinforcement learning, another machine learning technique, is employed to teach self-driving cars how to make complex decisions. In this approach, the car learns by interacting with its environment and receiving feedback based on its actions. Unlike supervised learning, where the algorithm is trained on labeled data, reinforcement learning uses a system of rewards and penalties to develop an optimal strategy for navigating various scenarios. Over time, the vehicle becomes proficient at making decisions that maximize rewards, such as avoiding obstacles and following traffic rules.

Predictive modeling is crucial for anticipating the behavior of other road users, a key aspect of safe autonomous driving. Machine learning algorithms analyze historical and real-time data to forecast the actions of pedestrians, cyclists, and other vehicles. For example, will a pedestrian step off the curb, or will another car change lanes? By predicting these movements, the self-driving car can adjust its course accordingly to prevent accidents. This aspect is particularly challenging as it requires dynamic decision-making based on constantly changing inputs.

Machine learning and AI algorithms also contribute to path planning and control. Once the vehicle has a clear understanding of its environment and the potential future states, it must decide the optimal path to take. Path planning algorithms, such as Rapidly-exploring Random Trees (RRT) and Probabilistic Roadmaps (PRM), help chart feasible routes by exploring all possible trajectories and selecting the safest, most efficient one. Following this, control algorithms ensure that the car follows the planned path by regulating its speed, steering, and braking.

The integration of deep learning with traditional AI approaches like rule-based systems and heuristic algorithms allows for more robust and adaptable autonomous driving systems. While rule-based systems can handle straightforward scenarios with clear rules and constraints, deep learning provides the flexibility to manage more complex, ambiguous situations that are harder to predefine. This hybrid approach maximizes the strengths of both methodologies.

Transfer learning, another significant development in machine learning, allows models trained in one domain to be applied to another. For instance, if a model is trained to recognize pedestrians in one city, it can be adapted to function in a different city or even for other types of objects. This capability significantly reduces the time and resources required for training models and makes the deployment of self-driving technology more scalable.

Ensuring the ethical and unbiased performance of AI algorithms is another important consideration. Biases in data can lead to unsafe or discriminatory decision-making by the vehicle. For instance, if the training data doesn't adequately represent all demographic groups, the car might not perform equally well in different settings, putting certain populations at greater risk. Ongoing efforts are made to diversify the training datasets and incorporate ethical guidelines into the algorithm development process to mitigate these risks.

Another layer of complexity arises from edge computing requirements, where data processing occurs closer to the data source (i.e., onboard the vehicle) rather than relying solely on centralized data centers. This decentralized approach minimizes latency, enabling real-time decision-making which is crucial for safety. AI chips specifically designed for edge computing environments optimize the performance of these algorithms while maintaining energy efficiency.

Training algorithms for autonomous vehicles is computationally intensive and requires vast amounts of data and processing power. Cloud-based simulations and virtual environments are used extensively to provide diverse and safe training grounds for these AI models. These simulated environments can replicate thousands of driving scenarios, including rare and dangerous situations that would be hard to encounter in real life. The synthesis of real-world and simulated data helps in building more resilient and comprehensive models.

Continuous improvement and learning are integral to the evolution of AI in self-driving cars. On-the-road learning allows vehicles to update and refine their algorithms based on real-time experiences. When a self-driving car encounters a new or challenging scenario, this data is shared across the fleet, allowing all vehicles within the network to learn from it. This collaborative learning approach accelerates the development cycle and enhances the overall safety and reliability of autonomous driving.

The synergy between machine learning, AI algorithms, and other technological advancements underscores the transformative impact of autonomous vehicles. As researchers continue to innovate and overcome existing challenges, it's clear that the future of transportation will be increasingly shaped by these intelligent systems. The relentless pursuit of safer, more efficient, and reliable self-driving technologies promises to redefine mobility, offering a glimpse into a future where roads are navigated by seamlessly intelligent machines.

Chapter 3:
The Evolution of AI in Transportation

The evolution of AI in transportation has been nothing short of transformative, marking a significant departure from traditional methods towards high-tech, efficient systems. From the experimental stages of early AI applications in the automotive industry to today's advanced self-driving algorithms, the journey has been fueled by groundbreaking research and innovative thinking. Noteworthy historical milestones, such as the development of the first autonomous vehicle prototypes in the 1980s, set the stage for today's rapid advancements. Influential researchers and pioneering companies alike have pushed the boundaries, integrating neural networks, computer vision, and sophisticated decision-making processes into modern transportation systems. As AI continues to evolve, its integration into various forms of transport promises to revolutionize how we move from one place to another, making travel safer, more efficient, and increasingly automated. This chapter delves into the profound shifts AI has driven in the transportation sector, exploring the pivotal moments and key players that have shaped this dynamic field.

Historical Milestones

The journey of AI in transportation is marked by numerous pivotal milestones. These moments have collectively shaped the trajectory of autonomous driving, transforming it from a futuristic concept into a tangible reality. One of the earliest and most significant milestones was the Defense Advanced Research Projects Agency (DARPA) Grand Challenge, held in 2004. This event invited teams to develop autonomous vehicles capable of navigating a 150-mile course through the Mojave Desert. Though none of the participating vehicles completed the course that year, it spurred tremendous advancements in the field and laid a foundational groundwork.

In 2005, the DARPA Grand Challenge saw remarkable progress. Stanford University's entry, "Stanley," not only completed the course but won the challenge by outperforming other competitors with its innovative AI algorithms and sensor fusion techniques. This victory showcased the potential of AI in navigating complex environments autonomously, proving that self-driving cars were no longer a distant dream.

Building on these successes, researchers and companies around the world began to focus more intensely on AI's role in transportation. In 2009, Google started its self-driving car project, later known as Waymo, which would become a leader in the industry. Utilizing a blend of sophisticated sensors, machine learning, and mapping, Google's initiative accelerated the pace of development and brought the concept of autonomous vehicles into the public eye. By 2012, Google revealed that its self-driving cars had logged over 300,000 miles without a single accident under autonomous control, highlighting AI's potential to improve road safety drastically.

Another significant milestone came in 2015, when Tesla introduced its Autopilot system, a semi-autonomous driving feature integrated into its vehicles. This system employed a combination of radar, cameras, and ultrasonic sensors to assist with driving tasks such as lane keeping, adaptive cruise control, and even some forms of automated lane changing. Tesla's integration of AI into consumer vehicles made autonomous driving technology accessible to the general public, marking a critical step toward widespread adoption.

Both the academic community and private sector continued to push boundaries. In 2016, Uber launched its first fleet of self-driving cars for passenger use in Pittsburgh, Pennsylvania. Though these vehicles operated under the supervision of human drivers, the move marked a substantial leap in the commercialization of self-driving technology. Around the same time, various tech giants and automotive companies, including General Motors, Ford, and NVIDIA, ramped up their autonomous vehicle programs, investing heavily in AI research and development.

The technological milestones have been complemented by legislative and regulatory milestones. In 2017, the United States House of Representatives passed the SELF DRIVE Act, aimed at accelerating the deployment of autonomous vehicles by setting federal standards for their operation. These regulations provided a clearer framework for the development and testing of self-driving cars, fostering a more conducive environment for innovation.

Global advancements also enriched the landscape. In 2018, Waymo launched the world's first commercial self-driving taxi service in Phoenix, Arizona. This service allowed customers to summon autonomous vehicles via a mobile app, operational without a human driver. This milestone not only demonstrated the feasibility of autonomous ride-hailing services but also underscored the maturity of AI technologies in real-world applications.

The following year, in 2019, AI-driven advancements continued to break new ground. Pony.ai, one of the leading autonomous vehicle startups, began offering free public rides in its fully autonomous vehicles in Guangzhou, China. This marked one of the first instances of public deployment of self-driving cars in a highly populated urban setting, providing valuable data and insights into the urban functionality of autonomous systems.

AI at the Wheel

By 2020, AI's role in transportation had reached an unprecedented level with the development of robust deep learning models and enhanced sensor technologies. Companies like Mobileye worked on creating AI systems capable of understanding complex traffic scenarios and making split-second decisions to ensure passenger safety. Innovations in AI algorithms and hardware have made it possible for autonomous vehicles to handle increasingly diverse driving conditions, from busy city streets to challenging weather environments.

In addition to commercial and academic efforts, collaborations between industry stakeholders and governments have also played a significant role. For instance, the European Commission launched the Horizon 2020 program with substantial funding allocated to AI and autonomous vehicle research. International partnerships have led to standardized protocols and safety measures, facilitating global advancements in the sector.

As we moved into 2021, AI-driven systems continued to improve, primarily through the use of advanced simulations and real-world testing. Companies like Zoox and Cruise began to unveil next-generation autonomous vehicles designed from the ground up, specifically for self-driving purposes. These vehicles incorporated the latest in AI, machine learning, and sensor technology to offer more reliable and efficient autonomous driving experiences.

In recent years, AI's integration into transportation has expanded beyond personal vehicles to include public and commercial transit. Autonomous buses and shuttles, such as those deployed by Navya, have started to operate in various cities worldwide. These innovations promise to revolutionize public transportation by increasing safety, reducing costs, and enhancing accessibility.

Looking at these historical milestones, it's clear that the development of AI in transportation has been both rapid and transformative. Each achievement has not only advanced the technology but also inspired further innovation and investment. The journey from DARPA challenges to commercial deployment highlights the collaborative efforts of researchers, tech companies, and policymakers in making autonomous driving a reality.

These milestones also set the stage for future advancements. With AI and machine learning continuing to evolve, we can expect even more sophisticated autonomous systems capable of revolutionizing how we think about transportation. As we look forward, the historical milestones serve as a testament to the power of innovation and the potential of AI to reshape our world.

Influential Researchers and Companies

The evolution of AI in transportation has been marked by groundbreaking advances, driven by visionaries and innovators across various fields. The fusion of artificial intelligence with the transportation sector can be credited to an interdisciplinary collaboration, where researchers and companies have played crucial roles. These key players have not only propelled the development of autonomous vehicles but have also shaped public perception and policy-making around self-driving technology.

Professor Sebastian Thrun, a pioneer in AI and robotics, is often regarded as one of the most influential figures in self-driving technology. Thrun led the Stanford Racing Team to victory in the 2005 DARPA Grand Challenge, a pivotal event in the history of autonomous vehicles. His work laid the theoretical and practical foundation for the development of modern self-driving cars. Later, he became co-founder of Google X, spearheading the Google Self-Driving Car Project, which eventually evolved into Waymo.

Waymo, a subsidiary of Alphabet Inc., stands as a testament to innovative corporate involvement in autonomous driving. Starting as

Google's ambitious project, Waymo has led numerous advancements in sensor fusion, machine learning, and real-world testing. Their approach integrates sophisticated algorithms with high-resolution maps, enabling a car to navigate seamlessly in complex environments. Waymo's continual progress signifies how corporate giants are crucial in steering the future of autonomous vehicles.

Dr. Amnon Shashua, co-founder and CEO of Mobileye, has been instrumental in bringing advanced driver-assistance systems (ADAS) to the market. Mobileye's technology includes computer vision systems that assist in collision prevention and lane-keeping, which are foundational for fully autonomous driving. Acquired by Intel, Mobileye continues to be a leader in providing AI-powered solutions for the automotive industry, influencing both consumer vehicles and commercial applications.

Tesla, under the leadership of Elon Musk, has become synonymous with cutting-edge automotive innovation. Tesla's Autopilot system represents one of the most prominently showcased examples of AI in personal vehicles. The company's ability to continuously update its software via over-the-air updates ensures that advancements in machine learning algorithms and sensor capabilities are swiftly integrated into the user experience. Tesla's robust data collection from millions of miles driven by its cars provides invaluable insights that push the boundaries of autonomous functionality.

Anthony Levandowski is another name associated with massive strides in autonomous technology. Known for his controversial yet significant contributions, Levandowski co-founded 510 Systems, which played a role in Google's self-driving car efforts. His work continued with the creation of Otto, a startup focused on autonomous trucks, which was quickly acquired by Uber. Despite legal battles surrounding his career, Levandowski's tech contributions underscore the rapid innovations catalyzed by determined researchers and entrepreneurs.

The academic realm has also been a vital battlefield for autonomous vehicle research. Laboratories at institutions like MIT, Carnegie Mellon University, and the University of Michigan have incubated numerous breakthroughs. At Carnegie Mellon, the Robotics Institute is a hub of innovation where researchers like Professor Raj Rajkumar have developed advanced navigation systems and machine learning techniques. Such institutions provide the foundational research that feeds directly into commercial applications.

Professor Fei-Fei Li from Stanford University has significantly influenced the field through her work in computer vision and large-scale visual recognition. Though her direct involvement with transportation AI might be limited, the applications of her research are broad and have implications across various autonomous systems, ensuring safer and more reliable vehicles. Her work on the ImageNet project revolutionized computer vision, facilitating more reliable object detection and scene understanding in self-driving cars.

Apart from academic and research institutions, various startups have disrupted the autonomous vehicle landscape. Aurora, co-founded by Chris Urmson, Sterling Anderson, and Drew Bagnell, is focused on providing the full-stack software to enable driverless cars. With significant industry backing and a wealth of experience among its founders, Aurora is poised to make substantial contributions toward achieving fully autonomous driving.

Another noteworthy entity is Nvidia, a company renowned for its GPU technology, which is integral to the computational needs of autonomous vehicles. Nvidia's Drive platform leverages deep learning to enable cars to interpret their surroundings and make real-time decisions. The adaptability and power of Nvidia's solutions have made them a favorite choice for numerous automotive manufacturers and tech startups alike.

One can't overlook Baidu, the Chinese tech giant, which has been a major player in the AI and autonomous vehicle space. Baidu's Apollo project serves as an open-source platform, offering a suite of autonomous driving capabilities to developers and manufacturers. This democratization of technology accelerates innovation and cross-industry collaboration, fostering a competitive yet cooperative environment that benefits the entire sector.

Additionally, Apple's secretive Project Titan indicates the company's keen interest in autonomous vehicles. While much of the work remains under wraps, Apple's significant investments and recruitment of top talent from the automotive and AI fields suggest that they are developing something transformative. The convergence of Apple's prowess in user-centered design and AI could yield groundbreaking results in autonomous driving technology.

In the realm of advanced simulation, companies like **Waymo and Nvidia** are utilizing virtual environments to test and validate driving algorithms. Simulation offers a risk-free setting to expose autonomous systems to a plethora of driving conditions and scenarios, from everyday situations to edge cases that would be dangerous to recreate in the real world. This ensures that self-driving cars are rigorously tested and better prepared for their deployment.

Finally, consider the role of international collaborations. Organizations such as the European New Car Assessment Programme (Euro NCAP) and Japan's New Energy and Industrial Technology Development Organization (NEDO) are pushing the boundaries of autonomous safety and innovation. These collaborations enable sharing of data, technology, and standards across borders, ensuring a unified and safer rollout of autonomous vehicles globally.

In summary, the amalgamation of influential researchers and forward-thinking companies is driving the rapid evolution of AI in transportation. From academic pioneers to tech giants and nimble startups, each contributor plays a vital role in overcoming technical challenges, addressing ethical and societal concerns, and bringing fully autonomous vehicles closer to reality. Their collective efforts underline a future where transportation is not just automated but also safer, more efficient, and accessible to all.

Chapter 4:
Sociocultural Impact of Autonomous Vehicles

The advent of autonomous vehicles is poised to reshape our social fabric in profound ways, influencing how we interact with one another and navigate public spaces. With self-driving cars eliminating the need for personal driving, the dynamic of urban landscapes will inevitably change as parking spaces diminish and streets become less congested. This transformation isn't purely physical; it's psychological as well. Public perception and social acceptance of these driverless wonders are crucial to their integration. Will communities embrace this technological leap, or will apprehensions about safety and job displacement hold us back? These questions underline the broader implications of autonomy in transportation, highlighting an evolving relationship between humans and machines that will redefine societal norms and expectations. Autonomous vehicles promise not just efficiency, but a cultural shift that could leave a lasting mark on our daily lives.

Changing Urban Landscapes

The advent of autonomous vehicles is set to reshape our cities in profound ways. This transformation goes beyond the novelty of seeing driverless cars on the streets; it touches how we plan our urban spaces, interact within them, and prioritize our infrastructure investments.

This section explores these shifts and the broader implications for our urban environments.

First and foremost, the reduction in the need for parking spaces may be one of the most visible changes. Cities devote vast amounts of real estate to parking lots and garages, often in prime locations. Autonomous vehicles can drop passengers at their destinations and move on to another task, park themselves in more remote locations, or simply keep moving. This potential reduction in parking demand opens up possibilities for repurposing central urban areas for public parks, gardens, and additional commercial and residential developments.

The flow and management of traffic might also experience dramatic changes. With improved efficiency and coordination among autonomous vehicles, traffic congestion could be significantly reduced. Cities could redesign streets to prioritize pedestrians and cyclists, creating safer and more enjoyable environments for those not in motorized vehicles. Narrower lanes, fewer traffic signals, and the integration of intelligent traffic management systems are all possibilities that could spring from a smoother flow of autonomous traffic.

Public transportation systems stand to be revolutionized as well. Self-driving buses and shuttles offer the promise of flexible, on-demand public transportation networks. This could make public transit more convenient and accessible, encouraging higher utilization rates and reducing the personal vehicle dependency that plagues many cities. Quieter, more efficient electric autonomous vehicles could also reduce noise and air pollution, contributing to cleaner and more livable urban spaces.

However, these urban transformations are not without their challenges. Cities will need to invest heavily in new infrastructure to support autonomous vehicles, such as advanced communication networks, charging stations, and sensors embedded throughout urban areas. These changes come with hefty price tags and require careful planning and coordination with both public and private sectors.

The redesign of roadways themselves will be a monumental task. Lanes might need to be adapted for the precise navigation of autonomous vehicles, and intersections could be reimagined to accommodate their unique capabilities. The process of updating urban infrastructure will necessitate significant investment, time, and a willingness to adapt to a rapidly changing technological landscape.

Another key aspect to consider is the potential social impact. Autonomous vehicles could democratize access to transportation for those who are currently underserved, such as the elderly, disabled, and underserved populations. This newfound mobility can foster greater inclusivity and participation in urban life. For instance, more people might be able to work or participate in social activities without relying on traditional, often unreliable, means of transportation.

The real estate market is poised for shifts as well. With less need for parking and a potential increase in livable public spaces, property values in urban centers might rise. Conversely, the demand for well-located parking garages and spaces could plummet, leading to a reevaluation of property value dynamics in cities. Moreover, residential areas might start appearing in places once considered too far from city centers due to the hassle of commuting, offering new opportunities for urban sprawl and development.

Moreover, urban planning must take into consideration the cybersecurity risks associated with autonomous vehicles. As more vehicles are controlled by complex algorithms and connected networks, ensuring the security of these systems becomes paramount to avoid potential threats. This concern extends to protecting critical infrastructure and sensitive data, necessitating robust safeguards and protocols to prevent cyberattacks.

The environmental implications also can't be overlooked. As more urban locales adopt autonomous vehicles, a shift towards electric or hybrid models is likely, further accelerating the reduction in greenhouse gas emissions. These changes could contribute to cleaner air and a lower carbon footprint, aligning with broader environmental goals and improving urban inhabitants' overall quality of life.

The community design might see more seamless integration of various transport modes. Bike-sharing programs, public transit, pedestrian pathways, and autonomous vehicle lanes could be harmoniously combined, facilitating easier transitions between transport modes and encouraging more sustainable living practices. Integrated mobility solutions can make cities more adaptable and capable of meeting diverse transportation needs.

Urban landscapes will also contend with ethical and regulatory considerations. Policymakers must navigate a complex landscape of privacy concerns, data usage regulations, and the equitable distribution of resources. Striking the right balance will be crucial to ensure that the benefits of autonomous vehicles are broadly shared and that no communities are left behind in the transition.

As autonomous vehicles become more ubiquitous, we might also see a lessening of the traditionally car-centric culture. Cities designed with people, rather than cars, at their center can foster more vibrant, active communities. Vibrant public spaces, increased social interactions, and a general shift towards pedestrian-friendly urban design are all potential outcomes of this profound transformation.

In sum, while the advent of autonomous vehicles promises remarkable changes to urban landscapes, it requires thoughtful consideration and careful planning. For this transformation to be successful, urban planners, policymakers, and technology developers will need to work together to create environments that are safe, sustainable, inclusive, and conducive to a modern way of life. These collaborations will help ensure that our cities are not only ready for the technological advancements of today but also adaptable to future innovations.

Social Acceptance and Public Perception

Autonomous vehicles (AVs) are one of the most transformative technologies of our time, promising to reshape the way we travel, work, and interact with our urban surroundings. However, the journey towards widespread adoption hinges significantly on social acceptance and public perception. Unlike other technological advancements, self-driving cars directly interface with public safety and everyday life, raising unprecedented questions and concerns among the general populace.

One of the first hurdles in the path to acceptance is trust. People need to believe that AVs are safe, reliable, and improve their quality of life. Initial surveys indicate a mix of excitement and trepidation among potential users. While younger generations and technology enthusiasts often express eagerness to embrace autonomous driving, older individuals and those unfamiliar with advanced technologies may harbor reservations.

Trust-building can be tied to several factors, including transparency in technological advancements, rigorous safety testing, and concrete evidence of improvement over time. High-profile incidents involving self-driving cars, such as fatal accidents during test drives, have cast a shadow over public sentiment. Media coverage tends to amplify these events, further complicating efforts to build trust.

Another complex issue is the psychological adjustment required to relinquish control to a machine. Many drivers are accustomed to the tactile and reactive nature of manual driving. The concept of an AI making split-second decisions on the road can be unsettling, even if statistics suggest machines could make fewer errors under the same circumstances. Companies developing these technologies need to invest heavily in public education campaigns to communicate the benefits and limitations clearly.

On the other hand, social narratives around safety and convenience can be powerful motivators for acceptance. Imagine a world where road fatalities plummet because AVs eliminate human error, or where commuting time turns into productive or leisure time because you're no longer behind the wheel. These positive scenarios need to be highlighted repeatedly to shift public perception towards the benefits rather than the risks.

In addition to individual comfort levels, societal norms and cultural attitudes also play a role in how AVs are perceived. Different regions may exhibit varying levels of acceptance based on local culture, existing infrastructure, and government policies. For instance, urban populations in tech-forward cities may be quicker to accept autonomous vehicles compared to rural communities where driving is ingrained into daily life and cultural identity.

The role of government and local institutions in fostering public acceptance cannot be overstated. Legislation that prioritizes safety, invests in infrastructure, and promotes equitable access can significantly affect public perception. Public pilot programs, such as autonomous shuttle services in specific urban districts, can serve as proving grounds for demonstrating the viability and safety of AV technology.

Media and popular culture also shape how autonomous vehicles are perceived. Films, television shows, and news coverage can either elevate or undermine trust in these technologies. Positive portrayals can demystify self-driving cars and position them as the logical evolution of transportation. Conversely, negative portrayals focusing on dystopian outcomes or malfunctions can deepen existing fears and misconceptions.

Social media platforms are emerging as important arenas for public discourse on AVs. Online communities can serve as echo chambers, where positive and negative opinions get amplified. Misinformation can spread quickly, making it crucial for manufacturers and policymakers to engage actively in these spaces, providing accurate information and addressing concerns directly.

Furthermore, early adopters and influencers can play a key role in shaping public opinion. When prominent individuals or institutions endorse a new technology, the public is more likely to view it positively. Testimonials, user reviews, and case studies showcasing real-world applications and benefits can also help in easing public skepticism.

Another dimension to consider is the ethical implications of AV technology. People are rightfully concerned about how these machines make decisions and whose safety is prioritized in critical situations. Transparent ethical guidelines and the inclusion of diverse perspectives in the decision-making process can help build public trust. Open dialogue and community engagement can make people feel more included in the technological transition.

Public perception also hinges on economic implications. There are fears about job displacement in sectors such as trucking and taxi services. However, highlighting new job opportunities and economic benefits that AVs bring can shift the narrative. As the technology creates new sectors and service models, public sentiment can evolve from apprehension to optimism.

Lastly, the speed of the transition to autonomous vehicles plays a role. Gradual implementation, starting with specific areas like public transportation or delivery services, can give society time to adjust.

Incremental updates and phased rollouts help mitigate risk and allow for continuous learning and adaptation.

Ultimately, the road to social acceptance and positive public perception is multi-faceted and requires a concerted effort from various stakeholders. Successful integration of autonomous vehicles into society will depend not only on technological advancements but also on how well these innovations are communicated, regulated, and perceived by the public.

Chapter 5:
Economic Implications of
Self-Driving Cars

The advent of self-driving cars is set to trigger profound economic shifts across multiple industries, challenging traditional norms while creating new opportunities. Automakers and tech companies are investing heavily in autonomous technology, promising growth in these sectors but also raising concerns about job displacement, particularly in driving-related professions. The ripple effects extend to associated industries such as insurance, logistics, and urban planning, potentially reshaping market dynamics and necessitating policy adaptations. On one hand, markets might surge with innovative services and products; on the other, challenges like regulatory compliance and ethical considerations will need to be addressed. Economically, the integration of autonomous vehicles could decrease operation costs, boost productivity, and even influence real estate values as commuting patterns evolve. Understanding these ramifications is crucial for stakeholders aiming to navigate this transformative landscape effectively.

Jobs and Employment Shifts

The advent of self-driving cars represents a significant technological breakthrough, one that brings with it a host of economic implications. Foremost among these is the potential transformation of the job market. The ripple effects of autonomous vehicles (AVs) on employment are inevitable, encompassing a wide array of industries from transportation to technology, affecting millions of workers globally.

Professional drivers are arguably the most immediately impacted group. Truck drivers, taxi drivers, and chauffeurs face the starkest reality. The U.S. Bureau of Labor Statistics reports that there are nearly 3.5 million professional truck drivers in the United States alone. These roles, traditionally reliant on human labor, are at risk of redundancy as AV technology advances. Initially, these vehicles might still require human oversight, but as technology matures, the need for a human driver could disappear altogether.

But it's not all doom and gloom for the workforce. While some jobs will be lost, new roles will inevitably emerge, reshaping the employment landscape. Much like previous industrial revolutions, jobs themed around the new technology will develop. IT support for AV systems, remote vehicle operators, and fleet managers for automated fleets are some roles that could see growth. Moreover, the development, programming, and maintenance of these sophisticated systems require a highly skilled technical workforce, creating opportunities in fields like AI, machine learning, and data science.

Beyond the direct roles, sectors such as insurance, legal, and urban planning will require professionals with expertise in the nuances of autonomous driving. Insurance agents and brokers will need to navigate the complexities introduced by self-driving cars, such as liability in the event of an accident. Legal professionals will be required to draft new regulations and navigate unchartered legislative waters to ensure public safety and fair commerce. Urban planners will have to rethink city spaces that accommodate both traditional and autonomous vehicles. These shifts can also lead to specialized training programs and educational curricula to prepare future workers for these novel challenges and opportunities.

Another significant consideration is the potential for geographic shifts in employment. As AV technology proliferates, tech companies and automakers might establish new hubs in areas previously untouched by the transportation industry. Conversely, traditional job centers heavily reliant on trucking and manually operated delivery services could experience a downturn, necessitating regional economic diversification efforts to bolster local economies.

Innovation in self-driving technology will give rise to secondary and tertiary job markets that service the primary autonomous vehicle industry. For example, with the increase in autonomous ride-sharing services, gig economy jobs might evolve from driving to tasks such as maintenance, cleaning, and customer service for these fleets. These roles, while different from traditional driving jobs, still offer avenues for employment and income.

Training and reskilling initiatives will be crucial. Outreach programs by governments and private sectors, aimed at retraining displaced workers, will play a significant role. Programs teaching new skills relevant to the autonomous driving industry — coding boot camps, AI workshops, and courses in data analytics — will become increasingly important. Providing access to education and skill development resources ensures that workers can transition into new roles created by this technological shift.

Additionally, demographic factors will influence how different groups experience these employment shifts. Younger workers, who generally adapt more quickly to new technologies, might find it easier to pivot to new job roles within the AV sector. In contrast, older workers might face more significant challenges, necessitating more robust support systems, from retraining programs to early retirement options.

Socioeconomic disparities could be both a cause and a consequence of this transition. Workers from economically disadvantaged backgrounds may find it more challenging to access opportunities for reskilling, exacerbating existing inequalities. Thus, ensuring equitable access to new job opportunities and training programs will be essential in mitigating these disparities.

The impact on labor unions and collective bargaining also warrants consideration. As the traditional job landscape shifts, unions that formerly represented truck drivers, taxi drivers, and other transport workers may need to adapt their mandates and strategies. Unions could play a pivotal role in negotiating fair transition schemes, advocating for reskilling programs, and ensuring that workers' rights are upheld amidst these changes.

Knowing that autonomous vehicles will dramatically alter supply chains, sectors reliant on extensive logistics operations will experience shifting employment demands. Warehouse and distribution center employees may find their roles evolving as companies adopt more sophisticated logistics systems driven by AV technology. Roles centered on the interface between human workers and automated logistics systems could become critical.

It's also worth noting that the broader economic landscape — including market growth and new opportunities created by AVs — will have a secondary effect on employment. As companies invest in AV technology, ancillary businesses such as software development firms, sensor manufacturers, and AV infrastructure developers will likewise expand, creating new jobs and contributing to economic growth.

In summary, the transition to autonomous vehicles has far-reaching implications for jobs and employment. While some traditional roles will be displaced, a myriad of new opportunities will emerge, reshaping the employment landscape. To navigate this shift successfully, the focus must be on proactive training, equitable access to new opportunities, and thoughtful policy development to support affected workers and communities.

Striking a balance among innovation, economic growth, and workforce transition will be crucial to harnessing the benefits of autonomous vehicles while mitigating the challenges they present. The road ahead is complex, but with proper planning and collaboration, a future with self-driving cars can also be a future with a dynamic and resilient workforce.

Market Growth and Opportunities

The autonomous vehicle market is poised for explosive growth over the next decade. This growth is driven by a combination of technological advancements, increased consumer interest, and substantial investments from both the private and public sectors. Self-driving cars represent a radical shift in how we envision transportation, offering significant opportunities for economic development, job creation, and innovation across multiple industries.

One of the most striking aspects of this market evolution is the diversity of sectors it impacts. From the automotive industry to technology companies specializing in AI, sensors, and telecommunications, a web of stakeholders benefits. Automakers like Tesla, Waymo, and traditional automotive giants such as Ford and GM are investing heavily in developing autonomous technology. Their commitment is evident through partnerships and collaborations with tech firms, leading to a competitive landscape that accelerates innovation.

According to market research, the global autonomous vehicle market was valued at approximately $54.23 billion in 2019 and is expected to reach $556.67 billion by 2026, growing at a compound annual growth rate (CAGR) of 39.47%. These figures highlight the enormity of the opportunity, making it one of the most attractive areas for investment and development in the coming years. Venture capitalists are particularly keen on this sector, pouring millions into startups focused on autonomous technology.

The commercialization of self-driving cars opens up new revenue streams. For instance, automakers transitioning from traditional manufacturing to autonomous vehicle production can capitalize on software updates, subscription services, and data monetization. It's not just about selling cars anymore; it's about creating a continuous revenue model around vehicle autonomy. Service providers offering autonomous taxi services or last-mile delivery solutions also stand to disrupt traditional logistics and public transportation markets.

While passenger cars often take the spotlight, commercial applications for autonomous vehicles (AVs) present substantial growth avenues. Autonomous trucks and delivery systems are already being tested and rolled out in pilot projects. Companies like Uber Freight and Amazon are exploring the efficiency gains and cost reductions associated with self-driving logistics. Freight transportation represents a multi-billion-dollar industry ripe for disruption, promising increased safety, lower operational costs, and reduced environmental impact.

In addition to the automotive and logistics sectors, ancillary industries also see potential in the rise of autonomous vehicles. Insurance companies, for instance, need to develop new policies and risk assessment models tailored to AVs. Infrastructure and urban planning will need to adapt, and companies specializing in smart cities and Internet of Things (IoT) systems can seize new business opportunities. Moreover, telecommunications firms will play a crucial role in ensuring the necessary 5G infrastructure for V2X (Vehicle-to-Everything) communication is in place.

Educational institutions and job training programs also have a tremendous opportunity here. As the market grows, so does the demand for skilled labor. Engineers, data scientists, and cybersecurity experts will be crucial in supporting and advancing autonomous vehicle technologies. Universities and training centers can offer specialized courses to prepare the next generation of professionals who will drive this industry forward.

Nevertheless, this burgeoning market isn't without its challenges. Regulatory hurdles, public skepticism, and ethical concerns about AI decision-making in life-and-death scenarios must be addressed. Policymakers have to strike a balance between fostering innovation and ensuring public safety. These challenges, however, create opportunities for companies that can navigate the regulatory landscape effectively. Firms specializing in compliance, legal consultancy, and ethical AI stand to benefit as advisors to the automotive and technology sectors.

Furthermore, international markets present varied opportunities. While the U.S. and parts of Western Europe are currently at the forefront, countries like China are catching up fast, investing heavily in autonomous technologies. Local conditions, such as urban congestion and government support, make different regions uniquely attractive for AV deployment. Companies aiming for global reach must tailor their strategies to meet the regulatory and cultural demands of each market, offering a plethora of opportunities for international business development and consultation.

One of the interesting opportunities lies in the secondary markets, such as fleet management and mobility services. As AV technology matures, businesses will look to modernize their fleets to include autonomous options. This creates a market for retrofitting existing vehicles with autonomous systems, a more cost-effective solution for many companies than purchasing new fleets outright. Additionally, ride-sharing platforms and startup mobility services can leverage autonomous technology to provide more efficient and cost-effective transportation solutions.

Public-private partnerships (PPPs) will likely play a crucial role in market expansion. Governments can provide the necessary infrastructure and regulatory frameworks, while private companies offer technological innovations. Such collaborations have already proven successful in pilot projects across the globe, from smart traffic light systems to dedicated AV lanes on highways. By working together, both sectors can accelerate the deployment of autonomous vehicles, creating economic benefits for all parties involved.

Finally, consumer acceptance is a critical factor that underpins market growth. As people become more familiar with autonomous technology through incremental implementations like advanced driver-assistance systems (ADAS), trust in full autonomy is expected to rise. Marketing strategies that educate the public about the benefits of self-driving cars—such as increased safety, convenience, and efficiency—can accelerate adoption rates. Studies show that younger generations are more open to the idea of sharing and autonomous driving, hinting at a future shift in transportation norms.

In summary, the market growth and opportunities surrounding self-driving cars are vast and multifaceted. The autonomous vehicle revolution promises not only to alter how we move but also to create a ripple effect that spans various industries, from high-tech and manufacturing to insurance and public policy. The road ahead is filled with potential, and the stakeholders who navigate these opportunities effectively stand to reap substantial rewards.

Chapter 6:
Ethics and Autonomous Driving

As autonomous driving technology speeds ahead, we're confronted with complex ethical questions that demand our attention. Self-driving cars must often make split-second decisions that carry significant moral weight, such as choosing between the lesser of two accidents. Beyond these moral dilemmas lies a web of regulatory and legal challenges that governments and companies need to untangle. Who is to be held accountable in the event of an accident? Furthermore, the algorithms guiding these vehicles aren't neutral; they carry the biases and limitations of their creators. Public trust hinges on transparent and fair development processes as well as legislative oversight. So while the promise of autonomous vehicles dazzles us with convenience and efficiency, navigating the ethical landscape requires us to tread thoughtfully, ensuring that technological advancements don't outpace our collective moral compass.

Moral Dilemmas

The advent of autonomous driving brings with it a plethora of moral and ethical quandaries. Autonomous vehicles (AVs) intrinsically involve decisions that could have life-or-death consequences. Delving into these moral dilemmas, we find ourselves navigating the intersection of technology and human values, a place where straightforward solutions are often elusive.

One of the most discussed ethical problems in AV technology is the "trolley problem." This classic moral dilemma challenges a decision-maker to choose between two harmful outcomes. Applied to AVs, it raises the question: Should a car prioritize the safety of its passengers over that of pedestrians? Imagine an AV faced with an unavoidable collision. The system must decide whether to swerve and endanger the life of its passenger or stay on course and risk hitting pedestrians. Programmers must pre-determine the vehicle's behavior in such scenarios, essentially embedding ethical decisions into the car's algorithms.

Adding dimensions to this dilemma, consider the diverse scenarios the AV might encounter. What if the pedestrians are children? What if swerving to save a group requires endangering a single, innocent bystander? How do cultural and societal values influence these decisions? Different cultures may prioritize different outcomes, making the development of a universal ethical framework for AVs exceptionally challenging.

The issue becomes more complicated when considering accountability. If an AV makes a decision that results in injury or death, who is held responsible? Is it the car's manufacturer, the software developers, or the owner of the car? In traditional vehicular accidents, identifying liability often falls to the driver. With AVs, the responsibility might diffuse across multiple stakeholders, complicating legal and insurance considerations.

Moreover, real-world application of these ethical decisions involves a countless number of variables. An AV must process real-time data and navigate continuously changing environments. This continuous need for ethical decision-making underlines the importance of robust, transparent algorithms capable of instantaneously resolving complex moral dilemmas.

This brings us to the ethical considerations surrounding the data used by AVs. Autonomous vehicles rely on extensive data sets to guide their decision-making processes. These data sets include demographic information, road conditions, and behavioral patterns of other drivers and pedestrians. How this data is collected, stored, and used presents its own set of moral questions. Is the public adequately informed about the data being gathered? Are there robust protections against misuse?

The concept of consent becomes central. To what extent are citizens willing and able to consent to their data being collected and utilized by AVs? Should there be an opt-in policy for data gathering? These questions are crucial as they intersect with privacy rights and the ethical use of personal information. Transparency from companies developing autonomous vehicles is essential in addressing these concerns.

Another layer of the moral fabric involves equity and accessibility. Autonomous vehicles have the potential to revolutionize transportation, making it more accessible to people with disabilities and improving the quality of life for many. However, these benefits must be universally accessible. If AV technology primarily caters to affluent individuals, it risks exacerbating existing social inequalities. Ensuring equitable access requires thoughtful policy and regulation.

The environmental ethics of autonomous driving also merit consideration. While AVs could significantly reduce emissions through improved efficiencies and optimized driving patterns, the production of these vehicles involves resource-intensive processes. How do we balance the environmental benefits against the ecological costs of manufacturing and maintaining AVs?

Contemplating these moral dilemmas without engaging in dystopian projections isn't simple. Public discourse often veers into alarmism, clouding objective analysis. It's essential to approach these ethical questions with a balanced perspective, recognizing both the immense potential and significant challenges AV technology poses.

On a more granular level, AVs introduce subtler moral dilemmas involving everyday human behavior. Consider a scenario where an AV must decide how aggressively to merge in heavy traffic. Should it prioritize punctuality, potentially leading to minor aggressive maneuvers, or prioritize safety, possibly inciting frustration among human drivers?

Even routine decisions, like whether to run a yellow traffic light, can become morally charged when placed in the hands of an autonomous system. These everyday ethical decisions require that AV designers instill a sense of judgment typically attributed to human drivers.

As a society, our collective answer to these dilemmas will shape the development and deployment of autonomous vehicles. Policymakers, technologists, ethicists, and the public must engage in ongoing dialogue to navigate these complexities carefully. We must recognize that while technology can advance, the core of these ethical dilemmas remains inherently human.

Educational initiatives can play a crucial role. Educating the public and industry professionals on the ethical dimensions of AVs can help foster informed discussions and better decision-making. Courses in ethics could become a standard part of the curriculum for engineers and data scientists working in this field, ensuring a grounding in the moral implications of their work.

Professional organizations and societies can also contribute to shaping ethical standards. By establishing ethical guidelines and holding member companies accountable, these organizations can help ensure that AV development adheres to shared moral principles. This collective effort will be essential in crafting solutions that are both ethically sound and widely accepted.

Ultimately, while the technology behind autonomous driving will continue to evolve, the ethical foundations we build now will guide its integration into society. Navigating these moral dilemmas requires a blend of technological innovation, thoughtful policy, and an unwavering commitment to human values. Only through such a holistic approach can we hope to harness the full potential of autonomous vehicles while minimizing their ethical risks.

Regulatory and Legal Challenges

The rise of autonomous vehicles (AVs) presents a unique set of regulatory and legal challenges. These challenges stem from the rapid pace of technological innovation outstripping the slower, more deliberate process of creating and implementing relevant laws and regulations. The landscape is complex and multifaceted, involving various stakeholders from government agencies to tech companies and the public.

One of the primary hurdles is creating a consistent regulatory framework that applies across different jurisdictions. Autonomous vehicles don't adhere to geographic boundaries; they are designed to travel anywhere. Currently, laws and regulations regarding AVs vary significantly from one region to another, even within the same country. This patchwork approach makes it difficult for manufacturers and developers to create systems that comply with all local regulations, stifling innovation and impeding the broader adoption of AVs.

In the United States, for example, each state has its own set of rules and guidelines. California requires autonomous cars to have a licensed driver behind the wheel, ready to take control in case of emergency, whereas states like Arizona have been more permissive in allowing fully driverless testing. This inconsistency can be a significant barrier for companies trying to bring AVs to market. It's not just about meeting safety requirements; it's also about complying with a myriad of different legal standards.

Internationally, the situation is even more complex. Countries such as Germany and Japan are pioneering detailed and stringent guidelines to govern the use of autonomous vehicles. Meanwhile, other nations lag in developing any significant regulatory oversight. This disparity creates a landscape where the speed of technological advancement in autonomous driving can be hindered by the slow pace of regulatory harmonization.

One of the most pressing issues is liability. When an autonomous vehicle is involved in an accident, who is at fault? In a traditional vehicle, the human driver is usually liable. However, an AV could shift that liability to the manufacturer, the software developer, or even the entity responsible for maintaining the vehicle's systems. This question of liability is not just academic; it carries significant financial, legal, and ethical implications. Companies are reluctant to assume full responsibility, yet the absence of a human driver makes the question unavoidable.

Some jurisdictions have begun to address these questions through new laws focused on AVs. However, these laws are often reactive rather than proactive, only coming into play after accidents occur. The Tesla Autopilot incidents, for example, have prompted many governments to reevaluate their regulatory frameworks. These incidents underscore the importance of establishing clear guidelines and responsibility before AVs become ubiquitous.

Another significant aspect pertains to data privacy and cybersecurity. Autonomous vehicles generate vast amounts of data, including information about passenger behavior, vehicle performance, and environmental conditions. This data is invaluable for improving the technology but also raises significant privacy concerns. Regulations like the General Data Protection Regulation (GDPR) in Europe aim to protect user data, yet they were not designed with AVs in mind. Adapting existing privacy regulations to cover the nuances of AV data is a task that regulators are only beginning to tackle.

Moreover, the cybersecurity of AVs is a critical regulatory concern. Hacking or tampering with an autonomous vehicle could have catastrophic consequences, from causing accidents to enabling malicious behavior such as vehicle theft or the use of an AV as a weapon. Consequently, regulators must ensure that companies adhere to strict cybersecurity standards, incorporating rigorous testing and validation protocols to protect vehicles from cyber threats. However, cybersecurity regulation is often several steps behind technological advancements, necessitating a dynamic regulatory approach that evolves as quickly as the technology itself.

Human safety and ethical considerations also present regulatory challenges. Decision-making algorithms in AVs may face situations where they must choose the "lesser of two evils." For instance, if a collision is unavoidable, should the vehicle prioritize the safety of its passengers over pedestrians? These ethical dilemmas require the creation of guidelines that govern how AVs should behave in critical situations. However, legislating ethics is incredibly complicated and fraught with moral ambiguity.

Insurance is another sector profoundly impacted by the advent of autonomous vehicles. Traditional insurance models rely heavily on driver behavior and accident history to assess risk and determine premiums. With AVs, these models need restructuring to account for different risk factors, such as system reliability, software faults, and the technological maturity of the vehicle. Insurers and regulatory bodies must collaborate to develop new financial frameworks that accommodate these changes.

Finally, public perception and acceptance are intricately linked with regulatory efforts. Effective regulations must ensure that AVs are safe and reliable, instilling public trust in the technology. Poorly designed or inconsistent regulatory frameworks can erode this trust, slowing down the adoption of AVs. Public advocacy groups, industry stakeholders, and government agencies need to work together to educate the populace about the benefits and limitations of autonomous driving, creating a transparent dialogue around the deployment and regulation of this emerging technology.

In summary, the regulatory and legal challenges surrounding autonomous vehicles are extensive and multifaceted. As these technologies continue to evolve, the regulatory landscape must adapt in tandem, balancing innovation with public safety and trust.

Chapter 7:
The Role of Big Data in Autonomous Driving

As autonomous driving technology advances, the sheer volume of data generated and processed becomes astronomical. Big Data plays a crucial role in this by capturing vast amounts of real-time information from an array of sensors, cameras, and GPS modules embedded in autonomous vehicles. This data is not just collected but meticulously analyzed to help the vehicle make split-second decisions, enhancing both safety and efficiency on the road. Machine learning models and AI algorithms rely heavily on this data to continuously learn and adapt, improving their accuracy and reliability over time. Big Data also supports predictive maintenance by identifying potential issues before they become critical, thereby reducing downtime and enhancing the vehicle's operational longevity. Moreover, the aggregated data helps in refining road safety measures and contributes to smarter city planning by providing insights into traffic patterns and infrastructure needs. Hence, Big Data isn't just a component but the very backbone of the intelligent systems that drive autonomous vehicles.

Data Collection Methods

In the quest to develop efficient, reliable, and safe autonomous vehicles, the collection of extensive and diverse data is paramount. Autonomous driving relies heavily on this data to train machine learning algorithms, refine AI systems, and ensure vehicles can handle a multitude of real-world scenarios. The methods used to collect this data are critical, encompassing a range of technologies and techniques designed to capture the complexities of driving environments.

One of the primary sources of data for autonomous vehicles is through sensor technology. These vehicles are equipped with a variety of sensors, including LiDAR, radar, cameras, and ultrasonic sensors. Each type of sensor has its advantages and plays a unique role in creating a comprehensive understanding of the vehicle's surroundings. For instance, LiDAR sensors offer precise distance measurements by emitting laser beams and calculating the time it takes for these beams to return. This helps create detailed 3D maps of the environment. In contrast, cameras capture visual data, providing images that are interpreted to detect objects, lane markings, and traffic signs.

Data collection doesn't end with what the sensors perceive in real-time. These vehicles also gather a plethora of information from their internal systems. GPS data provides accurate location tracking, while the car's speed, acceleration, and braking patterns offer insights into its performance and handling. This data is critical for understanding how the vehicle's characteristics interact with external conditions such as road quality and weather.

Another significant method of data collection is through simulation. Before autonomous vehicles are tested on real roads, researchers use sophisticated simulation tools to generate virtual driving scenarios. These scenarios replicate various traffic conditions, obstacles, and unexpected events that a vehicle might encounter. By running countless simulations, developers can expose AI systems to an extensive array of situations without the risks associated with physical testing. This virtual testing enriches the dataset and provides a safer, scalable means of gathering valuable data.

Crowdsourcing is also emerging as a vital data collection method. Companies like Tesla leverage their vast fleet of semi-autonomous vehicles to collect driving data from real-world conditions. These vehicles continually record information about their environments and driving behaviors, sending this data back to centralized servers. By analyzing terabytes of crowdsourced data, developers can gain insights into diverse driving conditions and improve the algorithms governing autonomous systems.

In addition to sensor data and crowdsourced information, High-Definition (HD) mapping plays a crucial role in data collection for autonomous vehicles. Traditional GPS maps aren't precise enough for the exacting demands of self-driving cars. HD maps provide highly detailed representations of the road infrastructure, including lane boundaries, curbs, and traffic signals. These maps are constantly updated with data collected from the sensors and other vehicles, ensuring that the autonomous systems have the most current information to navigate accurately.

Machine learning techniques further enhance data collection by enabling continuous improvement of AI models over time. As autonomous vehicles operate, they utilize a process known as continuous learning. This involves the vehicle's systems learning from each new mile driven, updating their algorithms with new data from diverse traffic scenarios. Continuous learning ensures that the AI systems evolve and adapt, becoming more adept at handling the complexities of real-world driving.

Data fusion is another critical aspect, amalgamating information from various sensors to create a cohesive understanding of the environment. Each sensor has limitations; for example, cameras might struggle in low-light conditions, while LiDAR can be hindered by heavy rain. By combining data from multiple sensors, autonomous systems can mitigate these weaknesses and achieve a more reliable perception of their surroundings. Data fusion algorithms synthesize inputs to enhance decision-making processes, contributing to the vehicle's overall safety and efficacy.

However, collecting data for autonomous driving isn't solely a technical challenge. Regulatory and privacy considerations play a critical role. Governments worldwide are enacting regulations to ensure data collected from autonomous vehicles is handled responsibly. This includes protocols for data encryption, anonymization, and storage, ensuring that sensitive information about drivers and pedestrians is protected. These regulatory frameworks are essential to gain public trust and facilitate the widespread adoption of autonomous vehicles.

The role of cloud computing in data collection can't be overlooked. Autonomous vehicles generate an enormous volume of data, far too much to be processed and stored locally. Cloud platforms offer scalable solutions for managing this data influx. They provide the computational power needed to analyze vast datasets, run complex algorithms, and update AI models in real-time. The seamless integration of cloud computing with autonomous systems ensures that vehicles constantly benefit from the latest advancements and insights derived from collective data analysis.

Moreover, interdisciplinary collaboration enhances data collection methodologies. Engineers, data scientists, AI experts, and automotive specialists work together to refine techniques and address challenges. Collaborative efforts result in innovative solutions, such as new sensor technologies or advanced data processing algorithms, pushing the boundaries of what autonomous vehicles can achieve. This synergy between various fields fosters progress and ensures that data collection methods are robust, efficient, and continually improving.

Ethical considerations are also intertwined with data collection practices. Developers must ensure that the data used to train AI systems is representative and unbiased. This involves deliberate efforts to collect data from diverse geographic locations, driving conditions, and demographic groups. Ensuring diversity in data collection helps prevent biases in AI models, promoting fairness and reliability in autonomous driving systems. Addressing these ethical challenges is imperative as the industry moves towards broader deployment.

Challenges persist in data collection for autonomous driving. The sheer volume of data can be overwhelming, requiring sophisticated storage and management solutions. Balancing data quality with quantity is another issue; while more data generally leads to better AI performance, it must be accurate and relevant to be useful. Moreover, the dynamic nature of road environments means that data quickly becomes outdated, necessitating continuous updates and validation.

Despite these challenges, advancements in data collection methods are paving the way for the future of autonomous driving. The integration of evolving technologies, regulatory frameworks, and collaborative efforts is creating a robust foundation for autonomous systems to thrive. As data collection techniques continue to improve, they will play an increasingly central role in refining the AI algorithms that power autonomous vehicles, ultimately creating safer and more reliable transportation solutions.

In summary, the methods used to collect data for autonomous vehicles are multifaceted, involving state-of-the-art sensor technologies, crowdsourcing, simulation, HD mapping, and continuous learning. Each method contributes unique insights and capabilities, together forming a comprehensive approach to understanding and navigating the complexities of real-world driving. Ongoing advancements in these methods are critical to the progress and success of autonomous driving initiatives, shaping the future of transportation and beyond.

Analytics and Real-Time Processing

As autonomous vehicles navigate through urban streets, rural backroads, and everything in between, the amount of data they generate is astonishing. This data isn't just a byproduct; it's the fuel that powers the advanced algorithms and decision-making systems critical to safe and efficient operation. The shift from manual to autonomous driving introduces a paradigm where real-time data processing and analytics become indispensable. In this chapter, we'll dive into how analytics and real-time processing shape the landscape of autonomous driving, focusing on the methods, technologies, and implications involved.

The essence of real-time processing in autonomous vehicles lies in its ability to make instantaneous decisions. Imagine a self-driving car approaching a busy intersection. The vehicle must process an array of inputs—traffic lights, surrounding cars, pedestrians, and even weather conditions—within milliseconds. With the help of sophisticated sensors like LIDAR, radar, and cameras, it collects raw data. This data must then be rapidly filtered, analyzed, and converted into actionable intelligence. These processes occur in microseconds, underscoring the imperative need for efficient analytics systems.

Data analytics in this context isn't just about processing speeds; accuracy and reliability are equally crucial. Real-time analytics enable autonomous vehicles to identify potential hazards and react appropriately. For instance, if a child suddenly runs into the street, the vehicle's system must immediately recognize the danger and initiate braking mechanisms. Delays or errors in data processing could have catastrophic consequences, emphasizing why this technology has to be flawless. Companies like NVIDIA and Intel are at the forefront, developing specialized processors designed to handle these specific demands.

One of the cornerstone technologies making this possible is edge computing. Unlike traditional computing, which may involve sending data to a distant server for processing, edge computing allows data to be processed at the "edge" of the network, closer to where it is generated. For autonomous driving, this means that the calculations needed to make real-time driving decisions can be performed directly within the vehicle. This significantly reduces latency and enhances the car's ability to make split-second decisions, leading to safer and more responsive driving experiences.

Edge computing isn't merely about responsiveness; it also addresses concerns over bandwidth and connectivity. Autonomous vehicles generate and require massive amounts of data. Relying solely on cloud-based processing would overwhelm network capabilities and introduce unacceptable delays. By distributing computational tasks to the edge, cars can function effectively even in areas with limited connectivity. Furthermore, this approach helps in managing privacy concerns, as sensitive data can be processed locally rather than being transmitted over potentially insecure networks.

To complement edge computing, Software Development Kits (SDKs) like ROS (Robot Operating System) provide a foundation for developing and deploying real-time analytics. These SDKs offer libraries and tools that simplify the implementation of complex algorithms, making it easier for developers to focus on refining the vehicle's analytic capabilities rather than reinventing the wheel. The collaborative nature of these development platforms ensures that advancements in analytics and real-time processing proliferate rapidly across the industry.

Real-time analytics also play a significant role in predictive maintenance, another crucial aspect of autonomous Driving. Sensors continually monitor the vehicle's components, predicting failures before they happen. By analyzing data patterns, the AI can forecast maintenance needs, thereby reducing downtime and enhancing the lifespan of the vehicle. This form of analytics not only ensures safety but also drives economic benefits by decreasing repair costs and optimizing operational efficiency.

Machine Learning (ML) and Deep Learning (DL) are pivotal in elevating real-time data processing to new heights. These advanced forms of AI enable vehicles to "learn" from data, improving their ability to identify objects, understand context, and make decisions over time. By training on vast datasets collected from millions of miles of driving, these systems evolve, becoming more adept at predicting and reacting to the myriad scenarios encountered on the road. Through reinforcement learning, for example, vehicles can continuously improve their performance, adapting to new environments and road conditions efficiently.

The sheer variety of data sources involved in autonomous driving introduces another layer of complexity. From geospatial data to traffic patterns and even social media feeds for real-time event updates, the volume and diversity of the data require powerful analytics engines. Real-time processing frameworks like Apache Kafka and Apache Spark have become instrumental in managing these voluminous data streams. These platforms offer robustness and scalability, enabling continuous data ingestion and processing without bottlenecks.

Vehicle-to-Everything (V2X) communication is another development that heavily relies on real-time analytics. This technology allows vehicles to interact with one another and with surrounding infrastructure, facilitating a more cohesive and coordinated traffic ecosystem. For example, vehicles can share data on road conditions, traffic jams, or accidents, allowing others to adjust their routes accordingly. This flow of information supports smarter decision-making, minimizing congestion and enhancing overall traffic safety.

Despite the advancements, challenges persist in real-time analytics for autonomous vehicles. One major hurdle is data integrity. Sensors can sometimes provide noisy or erroneous data, and distinguishing between reliable and faulty inputs is critical. Data fusion techniques, which combine information from multiple sensors to produce more accurate outcomes, are essential here. By cross-referencing data from different sources, vehicles can achieve a more holistic and accurate understanding of their environment, mitigating individual sensor errors.

Moreover, the ethical and legal implications of real-time processing can't be ignored. For instance, what happens when the analytics system faces a no-win scenario, such as having to choose between two forms of unavoidable harm? These moral dilemmas are ongoing areas of discussion and regulation, drawing interest from ethicists, lawmakers, and technologists alike. The transparency and explainability of AI decisions, facilitated by real-time analytics, are becoming more crucial as these systems take on greater roles in society.

As we look forward, the future of analytics and real-time processing in autonomous driving promises even more dynamic capabilities. Quantum computing, though in its nascent stages, holds potential for exponential improvements in data processing speeds and complexities. By leveraging principles of quantum mechanics, these systems could redefine the limits of what's achievable in real-time analytics, opening new frontiers in autonomous vehicle technology.

Additionally, advancements in 5G technology will offer faster, more reliable communication networks. This increase in connectivity will enable more sophisticated real-time analytics, enhancing the vehicle's ability to interact with its environment instantly. High-definition maps updated in real-time, augmented reality dashboards, and even more immediate predictive maintenance are just a few of the innovations that 5G could unlock.

In summary, the role of analytics and real-time processing in autonomous driving is multifaceted and indispensable. The integration of edge computing, advanced AI techniques, and innovative communication frameworks ensures that autonomous vehicles can navigate the complexities of real-world driving with unprecedented efficiency and safety. As technology evolves, so too will the capabilities of these systems, continually pushing the boundaries of what autonomous vehicles can achieve. This relentless progression underscores the transformative impact analytics and real-time processing will continue to have on the future of transportation.

Chapter 8:
Safety and Reliability in Self-Driving Cars

The quest for safety and reliability in self-driving cars is a focal point of their development, reflecting both technical challenges and ethical responsibilities. Ensuring these vehicles can navigate complex environments without human intervention requires sophisticated collision avoidance systems, informed by immense amounts of real-world data and rigorous testing protocols. Engineers and researchers employ layers of redundancies and fail-safes to handle unexpected scenarios, minimizing risks while enhancing trust. Besides technological advancements, achieving reliability across diverse driving conditions demands exhaustive validation processes, including simulations and real-world testing under varied conditions. This chapter delves into how these elements knit together to safeguard passengers and pedestrians alike, establishing a secure, dependable autonomous driving ecosystem.

Collision Avoidance Systems

The journey toward safe and reliable self-driving cars has led to many technological advancements, and collision avoidance systems (CAS) stand as a cornerstone in this evolution. CAS encompasses a suite of technologies designed to prevent or reduce the severity of collisions by alerting the driver to potential hazards or taking direct action to avoid an accident. For a vehicle to drive itself, these systems need to be exceptionally robust, offering reliability that meets or surpasses human capabilities.

One of the primary components of collision avoidance systems is the array of sensors deployed throughout the vehicle. LIDAR (Light Detection and Ranging), radar, and cameras are the most commonly used technologies. LIDAR provides a detailed 3D map of the environment, radar offers precise distance measurements and speed detection, while cameras capture visual data for object recognition.

The integration of these sensors allows the car to "see" its surroundings in great detail, providing a comprehensive view that human drivers could only dream of. Each sensor compensates for the limitations of the others, creating a multi-layered perception system that enhances overall accuracy and reliability.

Beyond hardware, the real magic happens in the software realm. Machine learning algorithms process vast amounts of data from the sensors to identify potential hazards such as other vehicles, pedestrians, or obstacles. These algorithms are trained on diverse datasets to improve their predictive capabilities, learning to recognize patterns and make decisions in real-time. The greater the diversity and volume of the training data, the better the system performs in unexpected scenarios.

Advanced driver-assistance systems (ADAS) form the foundation upon which collision avoidance systems are built. While ADAS functions like adaptive cruise control and lane-keeping assist are already prevalent in many modern vehicles, collision avoidance systems elevate these features to new levels of autonomy. In emergency scenarios, automated emergency braking can be the critical difference between a close call and a catastrophic collision.

One remarkable example of these systems in action is how they handle unexpected pedestrian crossings. By continuously scanning the environment and calculating the trajectories of moving objects, the system can predict when a pedestrian might step onto the road. If the situation becomes critical, the system can alert the driver or even apply the brakes automatically.

Another pivotal factor in the advancement of collision avoidance systems is the use of Vehicle-to-Everything (V2X) communication. V2X enables vehicles to communicate with each other and with infrastructure elements like traffic lights and road signs. This level of connectivity enhances the situational awareness of each vehicle, allowing for coordinated actions that can prevent accidents before they even develop. Imagine a scenario where a car equipped with V2X can "see" around a blind corner by communicating with another vehicle approaching from the opposite direction.

Of course, the reliability of collision avoidance systems isn't just about preventing accidents but also about ensuring consistent performance in variable conditions. This is where rigorous testing and validation protocols come into play. Self-driving cars undergo extensive simulations that replicate countless driving scenarios, from clear, sunny days to heavy rain or fog, ensuring that the collision avoidance systems perform optimally under all circumstances.

Moreover, real-world testing is indispensable. Companies developing autonomous vehicles often use controlled environments, like test tracks, to fine-tune their systems before deploying them on public roads. This phased approach allows for the incremental buildup of experience and data, further refining the algorithms and hardware integrations necessary for reliable collision avoidance.

The continual improvement of these systems also ushers in the need for updating and maintaining software post-deployment. Over-the-air (OTA) updates ensure that vehicles receive the latest enhancements without requiring visits to a service center. These updates can introduce new capabilities and optimize existing ones, ensuring that the systems keep pace with evolving technology and driving conditions.

Despite the technological marvels, collision avoidance systems aren't infallible. They must contend with a myriad of challenges, from inclement weather affecting sensor inputs to the unpredictability of human drivers sharing the road. Ensuring that these systems can handle such complexities is an ongoing quest that combines engineering prowess with insights from psychology and human behavior.

To address some of these challenges, redundancy and fail-safe mechanisms are built into collision avoidance systems. For example, multiple sensors might feed data into the system to ensure that if one sensor fails or provides incorrect information, the others can compensate. Additionally, algorithms continuously check and balance each other's decisions, minimizing the likelihood of false positives or missed detections.

Looking ahead, the development of quantum computing might further revolutionize collision avoidance systems. Quantum algorithms could process sensor data at unprecedented speeds, offering real-time insights and decision-making capabilities that exceed today's standards. This leap could make the already sophisticated systems even more reliable and efficient.

The promise of collision avoidance systems extends beyond individual safety; they have the potential to transform our broader transportation ecosystem. Reduced accident rates would lead to fewer traffic jams and lower healthcare costs related to vehicle accidents. Moreover, with fewer collisions, vehicle maintenance expenses would likely decrease, making transportation more affordable and efficient for everyone.

As we edge closer to fully autonomous vehicles becoming a common sight on our roads, collision avoidance systems will undoubtedly play an essential role. They are, after all, the keepers of safety, standing between us and the myriad uncertainties that characterize driving. Through technological innovation and continuous improvement, these systems promise to make our roads safer and our journeys more reliable, bringing us one step closer to a future where car accidents are an exception rather than the rule.

In conclusion, collision avoidance systems are a testament to how far we've come in automotive technology. Combining cutting-edge hardware with sophisticated software, these systems pave the way for safe and reliable self-driving cars. As we advance, these technologies will only become more refined, gradually changing the landscape of our roads and how we navigate them.

Validation and Testing Protocols

Ensuring the safety and reliability of self-driving cars hinges significantly on rigorous validation and testing protocols. In the realm of autonomous vehicles, these protocols serve not only as technical checklists but also as trust-building instruments. Given the high stakes involved—the promise of reducing traffic accidents, saving lives, and transforming transportation—validation and testing must be exhaustive and meticulous.

Validation starts with simulation environments. Before an autonomous vehicle ever touches asphalt, developers put it through millions of miles of simulated driving conditions. These virtual worlds are astonishingly detailed, allowing cars to experience a wide spectrum of scenarios, ranging from routine commutes to rare, hazardous situations. Advanced algorithms guide these simulations, continuously refining the vehicle's decision-making capabilities. This initial phase provides the groundwork for more tangible testing efforts to follow.

The next phase moves to closed-course testing. On controlled tracks, self-driving cars are subjected to various conditions that they might encounter on public roads. Engineers meticulously design these tracks to replicate urban intersections, highway merges, pedestrian crossings, and other real-world complexities. Importantly, this controlled environment allows for safe experimentation with edge cases, where unusual or extreme conditions might occur. The goal here is to stress-test the vehicle's systems, ensuring it reacts appropriately even under duress.

Once a level of confidence is established on closed courses, testing transitions to public roads but within geofenced areas. Here, the vehicles navigate real-world conditions, but only within pre-defined boundaries. This is crucial for observing how the vehicle interacts with human drivers and pedestrians, and how it responds to the unpredictability of real-life traffic. The data gathered from these tests is invaluable, feeding back into simulations and helping refine AI algorithms to improve performance and safety.

Advanced machine learning techniques play a vital role in these testing stages. Through continuous learning, the system adapts to new information and unforeseen scenarios. Self-driving cars rely heavily on neural networks to interpret sensor data, make decisions, and learn from each test run. The iterative process, comprising data collection, algorithm adjustment, and re-testing, ensures that the vehicles are in a state of constant improvement.

Scenario testing also extends to specific challenges like inclement weather conditions, diverse geographic terrains, and daytime and nighttime driving. Each of these factors presents unique complications that the autonomous system must master. For instance, inclement weather such as rain, snow, or fog significantly hinders sensor accuracy and vehicle performance. To counteract this, tests are conducted under varying weather conditions, allowing engineers to calibrate sensors and refine algorithms for optimal performance, irrespective of weather changes.

Another critical aspect is redundancy in hardware and software systems. Autonomous vehicles are equipped with multiple sensors—including LIDAR, radar, cameras, and ultrasonic sensors—each capable of providing overlapping data. This redundancy ensures that if one sensor fails, others can cover its duties, maintaining the integrity of the vehicle's decision-making process. Testing protocols must evaluate the performance of these redundant systems to verify that they function seamlessly and robustly.

Validation also includes rigorous cybersecurity testing. Self-driving cars are essentially networked computers on wheels and are thus susceptible to cyber-attacks. Testing protocols must encompass cybersecurity measures that preemptively identify vulnerabilities and implement protections. Engineers use penetration testing to simulate hacking attempts, ensuring that the vehicle's systems can withstand potential cyber threats. This aspect of validation is crucial in maintaining both the safety and the public trust in autonomous technologies.

Human-in-the-loop (HITL) testing is another layer adding to the robustness of validation protocols. Here, human operators monitor and, if necessary, intervene in the vehicle's operation during tests. This practice not only ensures real-time safety but also provides a human perspective on the system's decision-making capabilities. Data from HITL tests are instrumental in further fine-tuning the vehicle's behavior, aligning it closer to human driving norms.

Public road testing eventually extends beyond geofenced areas into more varied and unpredictable terrains. At this phase, self-driving cars must interact with the wider array of driving conditions and human behaviors prevalent across different regions. It's a more demanding test of the vehicle's reliability and safety, revealing how it performs under truly autonomous conditions. Stakeholders closely monitor these tests, ensuring that the vehicles meet stringent regulatory standards before they hit the consumer market.

Validation and testing programs also incorporate fail-safe mechanisms. Engineers build comprehensive checks into the system to handle unexpected failures gracefully. For instance, if a critical system malfunctions, the vehicle is designed to come to a safe stop. Testing protocols rigorously evaluate these fail-safes, confirming that they activate reliably under various conditions.

Moreover, validation isn't a one-time event but a continual process. As new updates and features are rolled out, they undergo the same rigorous testing and validation to ensure they integrate seamlessly into the existing framework. Continuous testing aligns with the evolving landscape of technologies and regulations, ensuring the vehicles remain at the cutting edge of safety and reliability.

Industry collaboration also plays a pivotal role in validation. Partnerships among automakers, tech companies, and regulatory bodies foster a unified approach to developing and validating autonomous vehicles. Shared data and standardized testing protocols ensure consistency and reliability across the industry. Such collaborations help set benchmarks for safety and performance, creating a transparent and accountable development ecosystem.

Regulatory authorities also mandate specific testing protocols before granting approval for public use. Agencies like the National Highway Traffic Safety Administration (NHTSA) and other international bodies have established guidelines that autonomous vehicles must meet to ensure public safety. These regulations form a vital part of the validation process, providing an additional layer of scrutiny and assurance.

Lastly, public trust in self-driving cars ultimately hinges on transparent communication of testing outcomes. Manufacturers must not only meet regulatory standards but also clearly convey the safety record and reliability of their vehicles to the public. Transparency about testing methodologies, results, and continuous improvements can foster consumer confidence and broader acceptance of autonomous technology.

In sum, validation and testing protocols are the cornerstone of ensuring the safety and reliability of self-driving cars. From simulation to public road testing, each phase builds upon the last, creating a comprehensive framework for evaluating and refining the technology. With continuous advancements and collaborative efforts, these protocols will evolve to meet the growing complexities and demands of autonomous driving, paving the way for a safer and more reliable transportation future.

Chapter 9:
Vehicle-to-Everything (V2X)
Communication

As we delve into the realm of Vehicle-to-Everything (V2X) communication, we encounter a world where cars don't just navigate on their own; they interact dynamically with their entire environment. V2X encompasses the spectrum of technologies that enable autonomous vehicles to communicate with other vehicles, infrastructure, and even pedestrians. This kind of connectivity is pivotal for enhancing road safety, reducing traffic congestion, and optimizing energy use. By facilitating high-speed data exchanges between smart cars and city infrastructure, V2X serves as the nervous system of a fully integrated, intelligent transportation network. The standards and performance metrics guiding this technology are continually evolving, pushing us closer to a reality where your morning commute is not just autonomous but also seamlessly connected to the world around you.

Infrastructure Requirements

The successful deployment of Vehicle-to-Everything (V2X) communication hinges on robust and adaptive infrastructure. V2X is about vehicles communicating with each other, traffic signals, road signs, and even pedestrians' smartphones. For this to happen, several infrastructure elements need to be in place, each playing a pivotal role in ensuring the seamless exchange of information.

First and foremost is the need for a comprehensive network of road sensors. These sensors, embedded in the asphalt or positioned along the roadside, are vital for collecting real-time data on traffic conditions, weather, and road hazards. This data is then relayed to V2X-equipped vehicles, helping them make informed decisions. The precision and reliability of these sensors are paramount, as inaccurate data could lead to hazardous scenarios on the road.

Roadside units (RSUs) are another critical component. These are essentially communication hubs placed along roads and highways. RSUs facilitate the exchange of information between vehicles and the broader traffic management system. They need to be positioned strategically to cover high-traffic areas while minimizing redundancy. RSUs must also be resilient to weather conditions and tampering, ensuring continuous operation.

Interoperability is a must for V2X communication to work effectively. Different vehicles and infrastructure components will utilize various communication protocols and standards. Establishing common communication standards that everyone must adhere to ensures that vehicles from different manufacturers can share information seamlessly. This necessity extends to international collaboration, as cross-border travel becomes more prevalent.

Reliable and high-speed internet connectivity is another cornerstone for V2X communication. Vehicles need real-time access to cloud-based services and databases to augment their decision-making processes. This reliance intensifies the need for widespread 5G implementation, given its low-latency and high-bandwidth attributes. While 4G networks currently support some V2X functionalities, 5G is expected to vastly improve the system's responsiveness and reliability.

Beyond technological hardware, there's also the need for sophisticated software systems. Centralized traffic management systems must be developed to process and analyze the massive amounts of data generated by V2X networks. Artificial Intelligence (AI) and machine learning algorithms will play a pivotal role in identifying patterns, predicting traffic flow, and suggesting optimal routes for autonomous vehicles.

Infrastructure isn't limited to just the physical hardware and software. Regulatory frameworks and policies must be established to govern the use of V2X technology. These policies will encompass data privacy concerns, system interoperability, and safety standards. Government bodies will need to work in close collaboration with technology companies, automotive manufacturers, and academic researchers to create a regulatory environment that fosters innovation while ensuring public safety.

Cybersecurity considerations cannot be overlooked. The interconnected nature of V2X systems makes them vulnerable to cyber-attacks that could have catastrophic consequences. Robust encryption methods, real-time threat detection, and secure communication protocols are essential to shield these systems from malicious entities. Continual updates and patches will be necessary to combat evolving cyber threats.

Public infrastructure enhancements are also needed. Traditional road signs and traffic signals must be upgraded to smart versions that can communicate with vehicles. Smart intersections, which can manage traffic flow dynamically based on real-time data, will be crucial in urban settings where traffic congestion is a significant issue.

The power grid is another area requiring attention. Many V2X infrastructure components, such as RSUs and smart traffic signals, will depend on a stable and reliable power supply. Integrating renewable energy sources and backup systems will ensure that these critical components remain operational even during power outages.

The deployment of V2X infrastructure also raises questions about the equity of access. Rural and underdeveloped areas may not have the resources to implement advanced V2X systems as quickly as urban centers. This disparity could lead to uneven safety and efficiency benefits, exacerbating existing inequalities. Policymakers must consider programs to support these areas, ensuring that they are not left behind in the transportation revolution.

Finally, public acceptance and collaboration will be key. Public and private sectors must engage in transparent communication, educating the population about the benefits and safety of V2X technology. Pilot programs, public consultations, and educational campaigns can help build trust and acceptance among the general public, fostering a smoother transition to this advanced infrastructure.

To summarize, the infrastructure requirements for V2X communication are extensive and multifaceted. Ensuring the success of V2X initiatives requires a concerted effort from multiple stakeholders, including policymakers, technology providers, and the public. By addressing these infrastructure needs comprehensively, we move closer to realizing a future where autonomous vehicles seamlessly integrate into our daily lives, enhancing safety, efficiency, and convenience on the roads.

Performance Metrics and Standards

The integration of Vehicle-to-Everything (V2X) communication represents a crucial step forward in the journey towards fully autonomous driving. V2X encompasses a broad array of communication types that link a vehicle with its environment, ranging from other vehicles (V2V), to infrastructure (V2I), and even to networks (V2N) and pedestrians (V2P). Because such a system has extensive implications for road safety, traffic efficiency, and energy optimization, establishing rigorous performance metrics and standards is essential.

To evaluate V2X communication, a comprehensive performance metric framework is employed to ensure the technology meets the stringent requirements necessary for real-world deployment. Key parameters such as latency, reliability, and coverage play prominent roles in dictating the success of V2X systems. Latency, for instance, refers to the time delay from the moment an information signal is sent from one entity until it's received by another. Lower latency is imperative for immediate hazard warnings, thereby reducing the likelihood of accidents.

Reliability, another critical performance metric, measures the consistency with which V2X communication delivers data within the prescribed time frame. In scenarios where vehicles traverse complex and densely populated urban environments, robust reliability ensures that vital information like traffic light signals or pedestrian crossing alerts is received without lapses. Failures in this area can lead to critical oversights and potentially hazardous situations.

Coverage is yet another essential parameter. V2X communication thrives on the extent to which it can provide service in varied geographic and infrastructural conditions. Urban areas with dense networks of high-rises, rural expanses with minimal infrastructure, and everything in between present unique challenges that need to be addressed to create a seamless V2X communication fabric. Effective coverage should guarantee that no vehicle is ever entirely outside the operational reach of V2X systems.

In addition to these primary metrics, factors such as data throughput and scalability are equally significant. Data throughput refers to the amount of data transmitted and received over the V2X network in a given time period. High throughput ensures that complex datasets, like HD maps or real-time video feeds, can be exchanged between vehicles and infrastructure without degradation. Scalability, on the other hand, ensures the system can handle an increasing number of connected devices without performance drops, a crucial attribute as the number of connected vehicles on the roads continues to rise.

Standards for V2X communication provide the regulatory framework within which these performance metrics operate. Globally, standardization bodies like the Institute of Electrical and Electronics Engineers (IEEE), the European Telecommunications Standards Institute (ETSI), and the 3rd Generation Partnership Project (3GPP) are actively collaborating to define these standards. The IEEE's 802.11p standard, known commonly as Dedicated Short Range Communication (DSRC), has been an early frontrunner in this arena.

The emergence of cellular vehicular-to-everything (C-V2X) communication further exemplifies this evolving standard landscape. Developed under the guidance of 3GPP, C-V2X leverages the advancements in cellular technology to offer improved reliability and coverage, especially in challenging environments. It benefits from large-scale cellular network deployments, integrating smoothly with the existing infrastructure and promising a forward-compatible path toward 5G.

Harmonizing these standards internationally is crucial to foster interoperability and avoid fragmentation. Interoperability ensures that vehicles from different manufacturers, operating in different regions, can communicate seamlessly with each other and with diverse infrastructures. This cross-functionality is essential for V2X to deliver on its promise of universal road safety and efficiency.

Real-world testing and validation of V2X systems serve as the crucibles where performance metrics and standards are put to the test. Large-scale pilot projects, such as those undertaken in smart cities like Singapore and San Francisco, offer valuable data and insights. These tests expose the technology to varied real-world conditions, revealing both its capabilities and vulnerabilities. Outcomes from these deployments feed back into the standardization process, ensuring continuous improvement.

Quantifiable performance metrics are invaluable here, providing objective criteria against which to assess V2X efficacy. Among these metrics, packet delivery ratio (PDR) has emerged as a fundamental measure. PDR quantifies the success rate of message delivery in the network and is particularly critical in safety-critical applications like collision avoidance. A high PDR indicates a robust and resilient V2X system.

Error rates form another key metric, particularly useful in assessing the reliability of communication under varying conditions. Metrics such as bit error rate (BER) and frame error rate (FER) help in identifying and mitigating sources of transmission errors, ensuring the integrity of data exchanged in the network. Low error rates correlate with high-quality communication links, which can make the difference between a minor inconvenience and a fatal accident.

Speed and accuracy of data processing within V2X systems also demand attention. The computational elements that handle data received from different entities must operate with precision and swiftness. More often than not, this involves complex machine learning algorithms capable of processing vast amounts of information in real time. The goal is to implement decision-making frameworks that can react swiftly to evolving road scenarios.

Security standards can't be overlooked either, as the open nature of V2X communication introduces potential vulnerabilities. Authentication and encryption protocols, such as those defined by IEEE 1609.2, are integral to ensuring that messages are exchanged between authenticated parties and remain tamper-proof. This aspect of performance metrics tackles the safety concerns posed by malicious intrusions and jamming attacks.

In sum, performance metrics and standards serve as the bedrock on which the success of V2X communication rests. They provide the benchmarks necessary to evaluate current technologies and guide future development. Through rigorous measurement and adherence to defined standards, V2X systems can evolve to meet the demands of safer, more efficient roadways. As this technology continues to mature, the continual refinement of performance metrics and standards will be indispensable in ensuring a harmonious integration of V2X capabilities into the broader tapestry of autonomous driving.

Chapter 10:
Energy Efficiency and Environmental Impact

The integration of autonomous vehicles into our transportation ecosystem is not just a technological leap but also a potential game-changer for environmental sustainability. These vehicles promise substantial reductions in emissions through optimized driving patterns and efficient route planning, leveraging advanced algorithms to minimize fuel consumption. Additionally, the push for sustainable materials and cutting-edge technologies in vehicle manufacturing is gaining traction, further bolstering the environmental benefits. By reducing the reliance on fossil fuels and incorporating renewable energy sources, the shift toward autonomous driving could significantly lower the carbon footprint of the transportation sector. The intersection of AI and eco-friendly innovations heralds a future where mobility and environmental stewardship go hand in hand, potentially altering the paradigm of energy consumption and environmental impact in profound ways.

Reducing Emissions

Reducing emissions stands at the core of the shifting paradigm toward autonomous driving. Transportation has long been a significant source of greenhouse gas emissions, contributing to climate change and poor air quality. The promise of self-driving cars includes more than just convenience and safety; it offers a roadmap to mitigate these environmental impacts significantly.

Electric vehicles (EVs) form the backbone of emission reduction strategies within autonomous transportation. By eliminating the need for internal combustion engines, EVs can operate without emitting greenhouse gases during use. When paired with renewable energy sources, the environmental benefits increase exponentially. However, it's not simply about the adoption of electric powertrains. Autonomous systems optimize EV performance through precise energy management, route planning, and efficient driving patterns, which collectively reduce energy use and emissions further.

AI-driven autonomous vehicles can effectively smooth traffic flow, reducing idling and stop-and-go behavior that leads to higher emissions. For example, machine learning algorithms can analyze traffic patterns in real time, predicting congestion points and rerouting vehicles to less congested routes. These strategies contribute to a more efficient use of fuel and energy, substantially cutting down emissions. In urban environments, this could translate to a significant reduction in CO_2 emissions and other harmful pollutants.

Moreover, autonomous vehicles have the potential to optimize car-sharing and ride-hailing services, leading to fewer vehicles on the road. With autonomous driving, shared vehicles could serve multiple users throughout the day, drastically reducing the need for private car ownership. This decrease in vehicle numbers could lead to lower collective emissions, as fewer cars would be produced, maintained, and eventually scrapped, reducing the overall environmental footprint.

The synergy between autonomous driving technology and the deployment of sustainable materials also plays a crucial role in emission reduction. Autonomous vehicles present opportunities for integrating lighter and more durable materials, such as advanced composites and bio-based substances. These materials contribute to better fuel efficiency and longer vehicle lifespans, further reducing emissions over time. For example, vehicles with lighter frames need less energy to move, thus emitting fewer harmful gases.

It's important to consider the lifecycle of autonomous vehicles, from production to disposal. AI can assist in designing efficient manufacturing processes that prioritize low-emission techniques. Autonomous technology may lead to advancements in recycling and material recovery, ensuring that end-of-life vehicles are repurposed more effectively. This closed-loop system can contribute significantly to minimizing the environmental impact of the automotive industry.

Autonomous vehicles also facilitate better implementation of V2X (Vehicle-to-Everything) communication systems, which can dramatically reduce emissions. These systems enable seamless communication between vehicles, traffic signals, and infrastructure, creating a more coordinated transportation network. By reducing instances of unnecessary braking and accelerating, AVs can ensure smoother traffic flow and decrease fuel consumption. This results in lower emissions in both urban and rural settings.

The geographical impact of emission reduction efforts can't be ignored. Different regions will benefit uniquely from the environmental efficiencies offered by autonomous vehicles. In densely populated urban areas, reducing emissions can alleviate smog and other air quality issues, leading to better public health outcomes. In rural areas, where distances between destinations are greater, improved fuel efficiency can make more significant reductions in overall emissions.

Public policy and government incentives are crucial in encouraging the adoption of emission-reducing autonomous technologies. Regulatory frameworks will need to evolve to support new energy-efficient standards for autonomous vehicles. Tax incentives, subsidies for electric and autonomous vehicle purchasers, and investments in renewable energy infrastructure can accelerate the transition toward a greener transportation future. The integration of these measures can help maximize the environmental benefits of autonomous driving technologies.

The integration of autonomous vehicles with renewable energy grids is another essential factor for emission reduction. Smart charging stations, powered by solar or wind energy, can be strategically placed to support the growing fleet of electric autonomous vehicles. Grid management algorithms can ensure that vehicles charge during periods of low grid demand or high sustainable energy output, further reducing the reliance on fossil fuels.

As technology evolves, the role of consumer behavior in emission reduction can't be overstated. Autonomous vehicles will come equipped with features that nudge drivers toward more eco-friendly habits. For instance, applications that rate the energy efficiency of driving routes or suggest carpooling options make it easier for individuals to choose lower-emission alternatives.

The collaboration between automotive manufacturers, AI developers, and environmental experts will accelerate the innovation needed to make autonomous vehicles viable climate solutions. Continued research and development in battery technology, energy-efficient computing, and sustainable manufacturing will drive down the emissions associated with autonomous driving. Partnerships with energy providers to ensure a green energy supply will also be critical.

In conclusion, reducing emissions through autonomous driving technology requires a multifaceted approach. The shift toward electric vehicles, smart traffic management, shared-vehicle models, sustainable materials, efficient lifecycle practices, V2X communication, and supportive policy frameworks all play indispensable roles. Together, these strategies pave the way for a significant reduction in transportation-related emissions, promising a cleaner, healthier future for our planet.

Sustainable Materials and Technologies

The automotive industry is undergoing a sea change, driven by the rise of autonomous driving technology. Yet, another profound transformation runs parallel—the shift towards sustainable materials and technologies. This evolution isn't just about making vehicles that drive themselves but doing so in a way that's kinder to our planet. Combining energy efficiency with a low environmental footprint isn't merely an option anymore; it's a prerequisite for the future of mobility.

First and foremost, let's talk about the key materials transforming automotive manufacturing. Metals like aluminum and high-strength steel are now incumbents in the quest for reducing vehicle weight. Lesser weight translates directly into higher energy efficiency, a critical factor for both electric and traditional vehicles. However, merely using lightweight metals isn't enough. Auto manufacturers are increasingly turning to composites and novel materials like carbon fiber, which offer exceptional strength-to-weight ratios, albeit at higher costs. But as technology advances, these materials are becoming more affordable, making them viable options for mass-market vehicles.

An equally groundbreaking development is the rise of bioplastics and recycled materials. Traditional plastics, derived from petroleum, contribute significantly to carbon emissions and environmental degradation. Bioplastics, made from renewable sources such as corn starch or sugarcane, reduce this impact considerably. Additionally, some automakers are experimenting with fully biodegradable materials for specific vehicle components. Imagine a future where car parts don't end up in landfills for centuries but decompose naturally without harming the environment.

Recycling isn't just limited to plastics. Metals, too, are seeing a new life thanks to sophisticated recycling methods. Companies are innovating to reclaim valuable metals like lithium, cobalt, and nickel from old batteries, reducing the need to mine, which is an environmentally taxing process. This circular economy approach to materials ensures that resources are used more efficiently, reducing waste and conserving natural resources. One noteworthy initiative involves closed-loop systems, where the end-of-life vehicle returns to the manufacturer to disassemble and reuse its raw materials in new vehicles.

It's imperative to highlight the advancements in battery technology as well. The traditional lithium-ion batteries, despite their ubiquity, pose significant environmental concerns due to their mining and disposal processes. Innovations such as solid-state batteries, which promise higher energy density and reduced risk of flammability, could revolutionize both the performance and environmental footprint of electric vehicles. Moreover, research into alternative materials like graphene and sodium-ion batteries continues, aiming to find solutions that are not only more efficient but also more sustainable.

Beyond materials, how these components come together in the manufacturing process also speaks volumes about sustainability. Modern factories are increasingly adopting green manufacturing practices. Powered by renewable energy sources like solar and wind, these facilities are designed to minimize waste and maximize efficiency. Processes such as 3D printing allow for precision manufacturing, reducing material waste. Digital twins, a virtual replica of physical assets, enable real-time monitoring and optimization of manufacturing processes, ensuring that resources are used judiciously.

The integration of smart technologies in manufacturing processes also ensures lower energy consumption. Internet of Things (IoT) devices offer unprecedented insights into the operational efficiency of manufacturing plants. These devices monitor everything from power usage to machine performance, identifying inefficiencies and suggesting corrective actions in real-time. Artificial Intelligence (AI) further augments these capabilities, using predictive analytics to foresee and mitigate energy waste before it happens.

However, sustainable materials and technologies are not limited to the components and their manufacturing. They span the entire lifecycle of the vehicle, from production and use to disposal. Lifecycle assessment (LCA) methodologies are becoming a standard practice for automakers. These methodologies assess the environmental impact of every stage of a vehicle's life, guiding more sustainable decision-making. For instance, LCA can help determine if the environmental cost of producing a new material is offset by its benefits during the use phase of the vehicle.

Another intriguing development is the concept of green tires. Traditional tires are a significant source of microplastic pollution, thanks to abrasion over time. Innovations in tire technology now include using sustainable materials like natural rubber and silica derived from rice husks. These new-age tires not only aim to reduce pollution but also enhance the vehicle's fuel efficiency by lowering rolling resistance.

Renewable energy isn't just revolutionizing manufacturing but is also pivotal in running these zero-emission autonomous cars. Renewable energy sources, particularly solar and wind, are increasingly harnessed to charge electric vehicles (EVs). Combining renewable energy with EVs creates a virtuous cycle: vehicles that not only emit no greenhouse gases but are also powered by clean energy. The rise of smart grids ensures that this renewable energy can be stored and distributed efficiently, aligning with peak demand times to reduce strain on the electrical grid.

It's also worth mentioning the infrastructure that supports these sustainable technologies. Charging stations equipped with solar panels and energy storage systems are popping up, particularly in forward-thinking cities. These stations help mitigate the grid's burden and ensure that the energy used for charging electric vehicles comes from sustainable sources. Moreover, wireless charging technology embedded in roads is being explored, promising even more seamless and eco-friendly energy solutions for the future.

Looking at the broader picture, the sustainability narrative ties into a larger ecological and ethical framework—whether it's fair labor practices in material sourcing or the ethical implications of resource extraction. Ethical considerations are driving corporations to source materials responsibly, ensuring that their supply chains do not exploit workers or deplete ecosystems irresponsibly. Certifications like Fair Trade and initiatives adhering to the U.N. Sustainable Development Goals are becoming benchmarks for responsible sourcing.

The push for sustainable technologies extends to the aftermarket as well. Products like remanufactured parts, which involve restoring used components to like-new condition, offer a sustainable alternative to buying new. This approach not only reduces waste but also lowers the carbon footprint associated with manufacturing new parts. Additionally, maintenance practices are evolving, with more emphasis on extending the life of vehicle components through predictive maintenance, facilitated by AI and IoT technologies.

Indeed, there's a growing consensus that the journey towards sustainable transportation won't be accomplished by incremental improvements alone. Radical innovations and a paradigm shift toward sustainability are essential. Policymakers, industry leaders, and consumers must come together to foster an ecosystem that rewards sustainability. Initiatives like carbon credits, green certifications, and stringent regulations on emissions and waste are pivotal in steering the industry towards greener pastures.

In conclusion, as we forge ahead into the autonomous driving era, embracing sustainable materials and technologies isn't just beneficial—it's essential. From light-weighting and recycling to green manufacturing and renewable energy integration, the strides being taken in this domain are nothing short of transformative. It's not just about driving the future but doing so responsibly, ensuring that technological advancements go hand-in-hand with environmental stewardship. The road ahead is long, but each innovation brings us one step closer to a sustainable, autonomous tomorrow.

Chapter 11:
The Future of Public Transportation

Public transportation stands on the verge of a transformative evolution, propelled by the advent of autonomous technologies. Imagine a seamless network of self-driving buses and shuttles that not only reduce traffic congestion but also improve accessibility and efficiency for urban dwellers. These autonomous vehicles could integrate effortlessly into existing systems, leveraging real-time data to optimize routes and schedules, thereby cutting down on waiting times and operational costs. As cities continue to grow and the demand for sustainable transit solutions escalates, the focus will undoubtedly shift toward creating smarter, more responsive public transportation infrastructures. In a not-so-distant future, the synergy of AI, machine learning, and advanced sensors could redefine the way we perceive and interact with public transport, making it safer, more reliable, and remarkably user-friendly.

Autonomous Buses and Shuttles

Autonomous buses and shuttles are quickly emerging as pivotal elements in the future landscape of public transportation. These self-driving vehicles promise to revolutionize how people commute, offering a blend of efficiency, safety, and accessibility that traditional modes of transportation can't match. As urban areas become more congested and the demand for greener, more sustainable transit solutions grows, autonomous buses and shuttles provide an innovative response to these challenges.

One of the primary advantages of autonomous buses and shuttles is their potential to significantly reduce traffic congestion. Unlike human drivers, autonomous vehicles can communicate with each other and with traffic management systems to optimize routes and speeds, ensuring smooth traffic flow. This real-time communication minimizes traffic jams and reduces the idle time vehicles spend at intersections. Moreover, these vehicles can operate around the clock without the need for driver shifts, thereby making public transport more available and reliable.

Safety is another critical factor where autonomous buses and shuttles outperform their human-operated counterparts. Equipped with advanced sensors, cameras, and AI algorithms, these vehicles can detect and respond to their surroundings with a level of precision that human drivers can't achieve. They can anticipate and avoid potential hazards, reduce the likelihood of accidents caused by human error, and ensure safer journeys for passengers. The integration of collision avoidance systems and real-time data processing capabilities plays a crucial role in enhancing safety standards.

A shift towards autonomous public transportation also represents a significant step towards environmental sustainability. Traditional buses and shuttles often run on fossil fuels and contribute to urban pollution. In contrast, most autonomous buses and shuttles are designed to be electric, offering a cleaner and more eco-friendly alternative. By reducing emissions and improving energy efficiency, these vehicles contribute to cleaner air and a reduced carbon footprint, aligning with broader environmental goals.

Autonomous buses and shuttles can also address accessibility issues in public transportation. In many cities, people with disabilities or those living in underserved communities face challenges in accessing reliable transit options. Autonomous shuttles can be designed with features that cater to these populations, such as wheelchair ramps, voice-activated controls, and flexible routing that can adapt to specific needs. This inclusivity not only enhances the quality of life for many but also offers a more equitable mode of transportation.

However, the integration of autonomous buses and shuttles into existing transportation systems is not without its challenges. One major hurdle is the need for robust infrastructure capable of supporting these advanced vehicles. This includes the development of smart roads equipped with sensors and communication technology, as well as dedicated lanes or areas where autonomous vehicles can operate without interference from traditional traffic. Cities must invest in this infrastructure to realize the full benefits of autonomous public transportation.

Public perception and acceptance also play a crucial role in the proliferation of autonomous buses and shuttles. While many people are excited by the prospect of futuristic transportation, others remain skeptical or fearful of putting their trust in AI-driven vehicles. Building public trust involves transparent communication about the technology's safety, reliability, and benefits, as well as providing opportunities for the public to experience these vehicles firsthand through pilot programs and demonstrations.

Moreover, the transition to autonomous public transportation will have significant economic implications. While the technology offers the potential to lower operational costs by eliminating the need for human drivers, it also raises concerns about job displacement in the transport sector. Policymakers and industry leaders must navigate these economic shifts carefully, potentially by reskilling programs for those affected and by creating new job opportunities in the maintenance and management of autonomous vehicle fleets.

A key aspect of the successful deployment of autonomous buses and shuttles is rigorous testing and validation. Unlike private autonomous cars that may operate with fewer passengers, public transport vehicles carry multiple people at any given time, necessitating higher standards of safety and reliability. These vehicles undergo extensive trials to ensure they can operate seamlessly in varied conditions, from densely populated urban centers to quieter suburban routes. Real-time processing of vast amounts of data helps in the prompt identification and rectification of any malfunctions or inefficiencies.

In many cities worldwide, pilot programs and trial runs of autonomous buses and shuttles are already underway. These initiatives provide valuable insights into both the strengths and weaknesses of the technology, offering a roadmap for future improvements. They serve as crucial steps towards widespread adoption, allowing stakeholders to fine-tune the technology and resolve potential issues before a full-scale rollout.

The integration of autonomous buses and shuttles into public transportation systems also necessitates collaboration between various stakeholders. Governments, transportation agencies, technology developers, and the public must all work together to create a cohesive and efficient system. Collaborative efforts can lead to the development of standardized regulations and protocols, ensuring the smooth operation and safety of these vehicles.

The future of transportation is undoubtedly autonomous, and buses and shuttles are at the forefront of this transformation. They epitomize the benefits of AI-driven technology in creating smarter, safer, and more sustainable urban mobility solutions. As cities continue to grow and evolve, autonomous buses and shuttles will play an integral role in shaping the transportation networks of tomorrow, making daily commutes not only more efficient but also more enjoyable and accessible for everyone.

In conclusion, the journey towards adopting autonomous buses and shuttles is complex but promising. The challenges are not insignificant, but neither are the potential rewards. With continued innovation, investment, and cooperation, autonomous public transportation will become a staple feature of modern cities, paving the way for a new era of mobility.

Integration into Existing Systems

The integration of autonomous vehicles into existing public transportation systems stands as one of the foremost challenges and opportunities in the transportation sector today. While the concept of self-driving cars often dominates discussions, the seamless incorporation of autonomous buses, shuttles, and support vehicles offers a transformative potential that could redefine urban mobility. This massive shift requires an intricate blend of technology, infrastructure, policy-making, and public buy-in. Multifaceted coordination between multiple stakeholders is imperative to ensure that this transition leads to practical, reliable, and inclusive transportation networks.

To begin with, the existing infrastructure must evolve to support autonomous vehicles effectively. Many cities already possess a foundation of public transportation systems, complete with dedicated lanes, traffic signals, and centralized control centers. However, these systems were designed for human-driven vehicles. Upgrades are necessary to facilitate the smooth operation of autonomous vehicles. Intelligent traffic signals equipped with Vehicle-to-Everything (V2X) communication capabilities can interact with autonomous buses in real time, enabling optimized traffic flow and reducing bottlenecks.

In addition to physical infrastructure, digital infrastructure plays a crucial role. Data hubs that can collect and analyze information in real time are vital. These hubs should be designed to handle vast amounts of data generated by sensors and AI algorithms in autonomous vehicles. This setup allows for real-time route optimization, predictive maintenance, and emergency response systems. With the integration of big data, traffic patterns can be analyzed, and inefficiencies in the current system can be identified and addressed.

Moreover, the inclusion of autonomous vehicles in public transportation isn't just a technological transition; it's a cultural one. Public perception and acceptance pose significant hurdles. Public perception will hinge on the perceived safety and reliability of these systems. Thus, rigorous testing and transparent communication with the public are essential. Pilot programs can serve as useful tools for both gathering data and demonstrating the technology's effectiveness and safety to the public.

From a regulatory perspective, establishing framework guidelines is indispensable. Current laws and regulations are primarily structured around human-driven vehicles. Autonomous technology introduces new challenges that require reevaluation and updating of these frameworks. Regulatory bodies need to set standards for safety, liability, and cybersecurity to ensure that autonomous systems are both reliable and secure. Collaboration between local governments, federal agencies, and tech developers is critical to crafting guidelines that are both practical and forward-thinking.

The involvement of private enterprises also significantly influences the pace and direction of integration. Tech giants, transportation companies, and startups alike bring innovative solutions and investments to the table. Public-private partnerships can provide the financial and technological resources needed to implement large-scale changes efficiently. For instance, automated fleet management systems developed by tech companies can be adapted for public transportation. These systems would use AI to manage schedules, allocate resources, and even predict and solve potential system failures before they occur.

Economic considerations can't be overlooked either. Implementing autonomous systems requires substantial upfront investments; however, the long-term benefits can be substantial. Autonomous buses and shuttles can lower operational costs through reduced labor expenses and enhanced fuel efficiency. Furthermore, they can increase accessibility, offering more consistent and extensive service routes that operate around the clock, thereby making public transportation a more appealing option for commuters.

In rural areas, the integration of autonomous transportation presents unique challenges and opportunities. Sparse populations and vast distances often make it economically unfeasible to offer frequent traditional public transportation services. Autonomous vehicles can fill this gap by providing on-demand transportation services, reducing the need for fixed-route buses that may often run empty. Tailoring autonomous solutions to meet the needs of both urban and rural environments ensures a more equitable distribution of resources and expands access to mobility for a larger segment of the population.

The social implications of integrating autonomous vehicles into public transport must also be addressed.

Skill requirements for public transportation employees will shift. Current drivers and conductors may find their roles evolving towards system oversight and customer assistance. Training programs will be necessary to equip these workers with the skills required for new job functions. Properly managed, this transition can minimize job displacement and even create new employment opportunities in tech and support services.

Public engagement initiatives can help address social fears and misconceptions. Community workshops, educational campaigns, and hands-on demonstrations can illustrate the benefits and safety features of autonomous systems, enhancing public trust. Companies and city administrations should prioritize transparency, actively engaging with the community to address concerns and provide regular updates on progress and safety measures, ensuring the transition feels inclusive and participatory.

Incorporating autonomous systems into public transportation also offers environmental benefits. Autonomous buses can be engineered with electric or hybrid powertrains, significantly reducing emissions compared to traditional diesel-powered buses. This transition aligns with broader sustainability goals and helps cities meet targets for reducing greenhouse gas emissions. Autonomous driving technologies optimize routes and driving patterns, further enhancing energy efficiency and reducing congestion-related emissions.

Finally, the user experience is an essential component of successful integration. Autonomous public transportation must be designed with the end-user in mind, ensuring accessibility, reliability, and ease of use. Features like automated boarding, real-time tracking, and seamless payment systems can greatly enhance user convenience. Additionally, designing these systems to be inclusive of people with disabilities and ensuring ease of use for senior citizens will be crucial in gaining wide acceptance.

The journey to integrating autonomous vehicles into existing public transportation systems is complex and multifaceted. However, with thoughtful design, robust policy frameworks, and active public engagement, it is possible to create a transportation future that is more efficient, equitable, and sustainable. The collaborative efforts between governments, private companies, and the public will ultimately determine the success and impact of this transformative technology.

As we move forward, continuous assessment and adaptation will be key in making autonomous public transportation a mainstream reality.

Chapter 12:
Autonomous Vehicles and
Ride-Sharing

As autonomous vehicles edge ever closer to mainstream adoption, they're set to revolutionize the ride-sharing industry in profound ways. Imagine a world where calling a ride means summoning a driverless car, ready to whisk you away efficiently and safely. The on-demand nature of ride-sharing pairs seamlessly with the capabilities of autonomous vehicles, offering increased flexibility and reduced wait times for consumers. From a business perspective, the economics of ride-sharing fleets are poised to transform, as companies can operate continuously without breaks, reduce labor costs, and optimize routes in real-time. This technological synergy not only promises more affordable and accessible transportation options but also hints at a future where urban landscapes fundamentally change, reducing the need for personal car ownership and rethinking city infrastructure. While challenges remain, particularly in regulatory and societal acceptance, the convergence of autonomous driving and ride-sharing offers a glimpse into a deeply interconnected, efficient transport ecosystem.

On-Demand Services

Autonomous vehicles are set to revolutionize the landscape of on-demand services, dramatically reshaping the transportation industry. From the shift in ride-sharing dynamics to the evolving business models, the technological leap includes significant societal changes. On-demand services enable users to request rides at their convenience, and autonomous driving technologies take this a step further by ensuring that service is consistent, safe, and available without human error. The core of this transformation lies in the interaction between artificial intelligence, ride-sharing platforms, and consumer needs.

One of the most noticeable impacts will be in the realm of ride-sharing. Current platforms like Uber and Lyft have already changed how people think about personal transportation. By using autonomous vehicles, these companies can offer even more reliable and cheaper services. With no need for a human driver, operational costs will decrease, translating into more affordable rides for consumers. Such developments enable the possibility of a ride-sharing environment where cars are continuously available, efficiently managed through advanced algorithms predicting and responding to real-time demands.

Consider how machine learning algorithms can optimize ride routes. These algorithms sift through massive amounts of data, including traffic patterns, historical ride data, and real-time road conditions. The result is a more efficient, faster service that matches supply and demand precisely. This real-time adaptability means fewer idle cars on the streets, lowering congestion and reducing the time passengers need to wait for their rides.

Moreover, the integration of autonomous vehicles into ride-sharing platforms is expected to herald a new age of convenience and accessibility. For instance, people living in areas with limited public transportation can benefit substantially. Access to reliable, on-demand transportation can improve quality of life by providing easier access to jobs, education, and healthcare services. It can also be a game-changer for individuals with disabilities, offering unparalleled mobility solutions.

The impact on urban planning and design will also be significant. As more people rely on autonomous ride-sharing rather than personal vehicle ownership, urban areas may see reduced demand for parking spaces. This reduction could free up valuable land for other uses, such as green spaces, pedestrian zones, or additional housing. Cities could be designed with fewer parking garages and more people-friendly environments, fostering a more livable and sustainable urban experience.

Meanwhile, the environmental advantages of on-demand autonomous services are hard to ignore. Electric autonomous fleets can reduce the carbon footprint of urban transportation. Autonomous technologies are inherently efficient, with algorithms that promote optimal driving behavior, such as smooth acceleration and deceleration, resulting in lower energy consumption. Moreover, multi-passenger ride-sharing options can further reduce the number of vehicles on the road, cutting down overall emissions.

From a business perspective, the shift to autonomous ride-sharing services encompasses diverse models, each with unique implications. Subscription-based services, where consumers pay a monthly fee for unlimited rides, could become widespread. Alternatively, per-ride pricing will continue to be popular, especially for occasional users. Regardless of the model, the industry is poised to see a significant reshaping as companies and consumers navigate this new landscape.

Real-time data analytics will underpin the success of on-demand autonomous services. Companies will need to collect and analyze data continuously to ensure their services meet market demands effectively. Predictive analytics can forecast demand spikes, allowing service providers to position their assets strategically. This predictive capability not only ensures high service levels but also maintains operational efficiency, creating a robust on-demand ecosystem.

The proliferation of ride-sharing services coupled with autonomous technologies raises questions about data privacy and cybersecurity. With vehicles continually collecting vast amounts of data, the security of this information is paramount. Companies must develop robust cybersecurity measures to protect user data from breaches and misuse. This necessity introduces regulatory challenges as governments and companies work together to create standards ensuring data protection.

So far, we've looked at the practical and societal implications. But on-demand autonomous services will also influence personal behavior and lifestyle choices. For many, the convenience of summoning a ride from their smartphone could make car ownership less attractive. This shift affects how people allocate their finances, possibly opting to spend money they would have used for owning and maintaining a car on other aspects of life.

The ripple effects extend to industries beyond transportation. Hospitality, healthcare, and entertainment sectors may see new opportunities arise. For example, hotels could offer integrated transport services, ensuring guests have immediate access to autonomous vehicles for their travel needs. Healthcare providers could partner with autonomous vehicle companies to offer rides for patients, enhancing access to medical facilities.

Ultimately, the adoption of on-demand services through autonomous vehicles is more than technological advancement; it's a societal shift. As AI continues to evolve and integrate seamlessly into everyday life, the collective impact on routines, urban landscapes, and economic models will continue to unfold. This dynamic transformation hinges on continuous innovation and strategic planning to fully realize its potential.

Business Models and Economics

Autonomous vehicles (AVs) are poised to revolutionize the ride-sharing industry, and their business models and economics form a crucial aspect of this transformation. Traditional ride-sharing services, like Uber and Lyft, depend heavily on human drivers and are subject to the limitations and unpredictabilities associated with human labor. As AV technology matures, it promises to eliminate these limitations, presenting both opportunities and challenges for stakeholders involved.

The economics of AV-based ride-sharing are fundamentally different from those of driver-operated models. Initially, the costs associated with AV technology—such as Research and Development (R&D), manufacturing, and regulatory compliance—are exorbitant. Companies must invest heavily in sensors, AI algorithms, and safety validation processes to ensure the vehicles can operate without human intervention safely. This up-front expense is a significant barrier to entry, but it is expected to decrease over time as technology advances and economies of scale come into play.

Despite the high initial costs, AVs offer substantial long-term savings. Operating costs are anticipated to be significantly reduced, as there is no need to compensate human drivers. This reduction in labor cost is particularly impactful in markets where labor constitutes a substantial portion of operational expenses. Additionally, AVs could offer improved fuel efficiency and reduced maintenance costs, thanks to optimized driving algorithms and less wear and tear compared to human-driven vehicles.

The shift towards AVs also introduces new revenue models for companies. Subscription-based models, pay-per-use, and dynamic pricing structures can be more effectively implemented in an autonomous environment. By leveraging big data, ride-sharing companies can offer personalized services, optimize fleet management, and dynamically adjust pricing based on real-time factors such as demand, traffic conditions, and user preferences.

For instance, a subscription model might allow users to pay a monthly fee for unlimited rides within a certain area, much like a public transit pass. Alternatively, ride-sharing companies could implement tiered pricing, offering different levels of service based on the type of AV used, the time of day, or the user's specific needs. The flexibility to innovate with pricing models affords companies a competitive edge and opens up new avenues for revenue generation.

Fleet management is another critical component of the AV business model. With human drivers, companies must coordinate shifts, handle labor disputes, and manage a workforce that is inherently unpredictable. Autonomous fleets, on the other hand, can be centrally controlled and optimized using sophisticated algorithms. These fleets can be deployed precisely where needed, balancing supply and demand more efficiently than traditional models. This optimization not only cuts costs but also enhances the user experience by reducing wait times and improving ride availability.

Additionally, AVs can exploit opportunities for downtime monetization. When not in use for ride-sharing, AVs could be employed for package deliveries, public utility services, or even mobile advertising platforms. These diversified use-cases provide additional revenue streams and maximize asset utilization, contributing to the overall profitability and sustainability of AV fleets.

Yet, the economic advantages of AVs are not without their trade-offs. The elimination of driver jobs is a significant concern, with potential social and economic repercussions. Job displacement might affect hundreds of thousands of individuals globally, necessitating measures such as retraining programs, social safety nets, and new employment opportunities within the AV ecosystem. Governments and companies must collaborate to mitigate these effects and ensure a just transition for the workforce.

Furthermore, the regulatory landscape for AVs is still evolving. Compliance with varying national and local regulations can incur additional costs and complexities. Companies must navigate this dynamic environment carefully, investing in lobbying efforts and maintaining flexibility to adapt to new laws and guidelines. The regulatory framework will play a critical role in shaping the economic viability of AV-based ride-sharing services.

Investment in infrastructure is another economic consideration. AVs require high-quality, well-maintained roads, advanced traffic management systems, and efficient charging or fueling stations. The development and upkeep of this infrastructure involve substantial public and private investments, and the costs must be justified by the anticipated benefits of improved safety, efficiency, and reduced environmental impact. Public-private partnerships could be crucial in balancing these costs and ensuring the infrastructure meets the needs of AV technologies.

On the consumer side, the economics of AV ride-sharing will influence adoption rates. While AVs promise lower per-mile costs compared to traditional ride-sharing, consumers will weigh these savings against concerns over safety, privacy, and the novelty of the technology. Effective marketing strategies and positive user experiences are essential to building trust and encouraging widespread adoption. Incentives such as introductory discounts or loyalty programs could also play a role in attracting early adopters and driving market penetration.

Looking ahead, the interplay between competition and collaboration will shape the economic landscape of AV-based ride-sharing. Companies that develop proprietary technologies may initially have a competitive advantage, but widespread adoption might hinge on standardization and interoperability. Collaborations between tech firms, automotive manufacturers, and ride-sharing companies could foster an ecosystem where innovations proliferate, costs decrease, and consumer choice expands. While competition spurs innovation, strategic alliances ensure the rapid and efficient deployment of AV technologies.

In summary, the business models and economics of autonomous vehicle ride-sharing present a complex tapestry of opportunities and challenges. While upfront investments are substantial, the promise of reduced operating costs, new revenue models, and optimized fleet management holds great potential for profitability. Addressing the social implications of job displacement, navigating regulatory hurdles, investing in infrastructure, and fostering consumer trust are crucial to realizing the economic benefits of this transformative technology. As the industry evolves, the balance between competition and collaboration will be key to shaping a sustainable and prosperous future for AV-based ride-sharing services.

Chapter 13:
The Role of Governments and Policy-Making

Governments play a crucial role in setting the stage for the widespread adoption of autonomous vehicles, acting as both regulator and collaborator. Policy-making involves establishing robust regulations and licensing criteria that ensure safety and reliability without stifling innovation. By collaborating internationally, governments can align standards, fostering a cohesive global framework that supports seamless cross-border operations. Public sector initiatives also emphasize infrastructure upgrades, cybersecurity, and privacy protections, ensuring that autonomous systems integrate smoothly with existing urban and rural landscapes. Through a balanced approach, policy-makers can foster an environment where technological advancements in self-driving cars flourish while addressing societal and ethical concerns, ultimately guiding the transformation of the transportation ecosystem.

Regulation and Licensing

The advent of autonomous vehicles (AVs) has prompted governments globally to reevaluate existing regulatory frameworks and develop new policies. As self-driving technology continues to evolve, the role of regulation and licensing becomes increasingly crucial to ensuring public safety and fostering innovation. Balancing these objectives presents policymakers with significant challenges.

In many jurisdictions, regulatory bodies have initiated pilot programs to test AVs under controlled conditions. These programs are designed to gather data on the safety, performance, and social impact of autonomous vehicles. This data is critical for developing comprehensive regulations that address the unique aspects of AV technology. For instance, new standards for vehicle safety certifications are being drafted to include specific requirements for the sensors and algorithms that drive these vehicles. It's a delicate balancing act between promoting technological progress and safeguarding public interest.

One key aspect of regulating autonomous vehicles involves updating driver licensing requirements. Traditional driver's licenses assume a human operator is in control, but AVs challenge this notion. Legislators need to define what constitutes "control" when a vehicle can drive itself. Should a person still be required to hold a conventional driver's license to oversee the vehicle, or would a new type of certification be more appropriate? Some regions are exploring the idea of "operator licenses" that certify individuals to monitor and intervene in AV systems when necessary.

Furthermore, there are ongoing discussions about the liability in accidents involving autonomous vehicles. Current regulations often place the burden on human drivers, but the dynamic changes when the vehicle itself makes decisions. This shift necessitates amendments in liability laws to hold manufacturers and software developers accountable to some extent. Insurance policies are also adapting, with new models being devised to incorporate the technological reliability and maintenance of AVs into the risk assessment.

International collaboration is another vital element in the regulation and licensing of autonomous vehicles. Different countries are at varying stages of AV adoption, and their regulatory practices differ accordingly. Harmonizing these practices is essential for fostering global innovation and enabling cross-border travel for AVs. Organizations such as the United Nations Economic Commission for Europe (UNECE) are working to establish international standards that could serve as a baseline for national regulations. Aligning these standards can help reduce the regulatory fragmentation that often hampers technological advancement.

At the domestic level, states and municipalities are also getting involved in the regulatory process, adding another layer of complexity. Various states in the USA, for example, have enacted divergent regulations regarding the testing and deployment of AVs. While some states have embraced more flexible guidelines to lure tech companies, others have opted for stringent regulations prioritizing public safety. This patchwork of policies necessitates a coordinated approach to create a cohesive national framework that can be adapted locally based on specific needs and conditions.

Moreover, the regulations extend beyond mere licensing and safety standards. They also encompass ethical considerations inherent in AI-driven decision-making. Autonomous vehicles must often make split-second decisions that could have moral ramifications—choosing between the lesser of two evils in accident scenarios, for example. Policymakers are now tasked with framing guidelines that dictate the ethical programming of these machines. This involves extensive public consultation and input from various stakeholders, including ethicists, technologists, and the general populace.

Transparency in the regulatory process cannot be overstated. Public trust in autonomous driving technology hinges on the perceived fairness and comprehensibility of the regulations that govern it. Regulatory bodies must strive to make their processes as transparent as possible, offering clear explanations for the rules they enact and the methods used to enforce them. Engaging with the community through public forums and feedback sessions can significantly improve the overall acceptance of AV regulations.

Economic implications also play a crucial role in shaping regulations. Autonomous driving technology promises to revolutionize various sectors, including public transportation, logistics, and personal mobility. Policymakers must consider how regulations can facilitate these transformations while minimizing potential disruptions. For example, in public transportation, regulations might need to encompass the integration of autonomous buses and shuttles within existing transit systems. This requires a collaborative effort between transportation authorities, technology providers, and urban planners to ensure seamless and efficient service delivery.

In conclusion, as the landscape of autonomous driving continues to evolve, so too must the regulatory frameworks that govern it. The challenge for policymakers is to create regulations that are flexible enough to adapt to rapid technological advancements while robust enough to ensure public safety and trust. Successful regulation and licensing of autonomous vehicles will be characterized by collaboration, transparency, and a forward-thinking approach, setting the stage for a future where autonomous driving becomes an integral part of everyday life.

International Collaboration and Standards

As autonomous driving technology advances, the role of international collaboration and standards becomes increasingly essential. Governments worldwide recognize that unified standards are critical for the development and implementation of autonomous vehicles (AVs). Without harmonized regulations and cooperation among nations, the deployment of these technologies could be fragmented and chaotic. Therefore, fostering global partnerships is fundamental for creating a cohesive framework that enables safe and efficient autonomous driving.

One of the primary arenas for international collaboration is in the formulation of technical standards. Technical standards ensure that autonomous vehicles, irrespective of their country of origin, can operate seamlessly and safely across different jurisdictions. Organizations like the International Organization for Standardization (ISO) and the Society of Automotive Engineers (SAE) are leading efforts to develop comprehensive standards for AVs. These standards cover a wide range of aspects, from vehicle-to-everything (V2X) communication protocols to cybersecurity measures, forming the backbone of the global autonomous driving ecosystem.

Collaborative standard-setting also promotes interoperability, which is crucial for the widespread adoption of autonomous vehicles. For instance, when AVs from different manufacturers adhere to the same communication protocols, it ensures that they can "talk" to each other effectively. This inter-vehicle communication is vital for preventing accidents and improving traffic flow. Such interoperability is achieved through regulatory alignments and agreements facilitated by international bodies, ensuring a unified global landscape for autonomous vehicles.

Moreover, international partnerships extend beyond just technical standards; they also focus on regulatory frameworks. Countries must align their safety regulations, testing procedures, and liability laws to create a conducive environment for AVs. Collaborative platforms such as the United Nations Economic Commission for Europe (UNECE) provide a stage where different nations can negotiate and codify regulations that harmonize safety and performance criteria. Such efforts prevent regulatory patchwork, which can be a significant impediment to the global rollout of AV technologies.

A vital component of these international collaborations is data sharing. Autonomous vehicles generate massive amounts of data, which can be invaluable for improving safety, performance, and efficiency. Cross-border data sharing initiatives can accelerate technological advancements by enabling researchers and developers to access diverse datasets. Joint research programs funded and supported by multiple nations enhance the quality and applicability of the solutions developed. For instance, the European Union's Horizon 2020 program has funded numerous projects focused on autonomous driving, including partnerships with non-EU countries, fostering a broader collaborative research environment.

Safety is another critical area where international collaboration plays a pivotal role. Setting universal safety standards ensures that autonomous vehicles meet consistent safety benchmarks, regardless of where they are manufactured or operated. Global initiatives, like the Global Forum for Road Traffic Safety (WP.1), work to establish these safety standards, providing a benchmark that all countries can adopt. This global alignment is crucial for public trust and acceptance of autonomous vehicles, as it reassures the public that AVs are held to stringent safety standards globally.

International collaboration is also crucial for addressing the cybersecurity risks associated with autonomous vehicles. Cyberattacks on AVs could have catastrophic consequences, making robust cybersecurity standards indispensable. Cross-border cooperation in cybersecurity helps develop comprehensive strategies to protect AVs from potential threats. By sharing knowledge and resources, nations can develop more effective defenses against cyber threats, ensuring the safe operation of autonomous vehicles. Initiatives like the United Nations Global Cybersecurity Index foster such international cooperation, highlighting best practices and areas for improvement.

The environmental impact of autonomous driving is another area where international standards can make a significant difference. By setting unified environmental standards, countries can ensure that the deployment of AVs contributes to global sustainability goals. For instance, international agreements can mandate the use of zero-emission technologies and sustainable materials in AV manufacturing. Such standards not only mitigate the environmental impact but also encourage the automotive industry to innovate in creating greener solutions.

Testing and validation of autonomous vehicles are other domains benefiting from international collaboration. Standardized testing protocols ensure that AVs undergo rigorous evaluations before they hit the road. Governments and international organizations work together to develop these protocols, encompassing various scenarios that AVs might encounter globally. This standardized approach simplifies the testing process for manufacturers and assures the public of the vehicles' reliability and safety.

On the policy front, international collaboration helps in streamlining insurance and liability frameworks for autonomous vehicles. Different countries might have varied legal perspectives on liability in the event of an AV-related accident. Harmonizing these legal frameworks ensures clarity and consistency, making it easier for manufacturers, insurance companies, and consumers to navigate the complexities of AV-related incidents. Platforms like the International Transport Forum (ITF) facilitate discussions and agreements on these critical issues, promoting a more cohesive global policy environment for AVs.

Another key aspect of international collaboration is addressing ethical considerations in autonomous driving. As AVs make decisions that could affect human lives, it is essential to align ethical standards globally. International dialogue helps in developing ethical frameworks that guide the decision-making processes of autonomous systems. These discussions ensure that AVs operate in a manner that is ethically acceptable across different cultures and societies, fostering global trust in the technology.

International collaboration is also instrumental in public education and awareness about autonomous driving. Coordinated efforts can help disseminate accurate information about AVs, dispelling myths and addressing public concerns. Joint campaigns by governments and international organizations can promote a better understanding of the benefits and limitations of autonomous driving, enhancing public acceptance and support.

In conclusion, the success of autonomous driving technology heavily relies on robust international collaboration and standards. By working together to develop unified technical standards, regulatory frameworks, safety protocols, cybersecurity measures, and ethical guidelines, nations can create a cohesive and supportive environment for AVs. This collaboration not only accelerates technological advancements but also ensures that the deployment of autonomous vehicles is safe, ethical, and beneficial for society as a whole. Through these concerted efforts, the global community can harness the full potential of autonomous driving technology, paving the way for a smarter, safer, and more sustainable future in transportation.

Chapter 14:
Cybersecurity in Autonomous Vehicles

As autonomous vehicles become increasingly integrated into our daily lives, cybersecurity emerges as one of the most critical challenges to address. Hackers targeting these sophisticated systems could cause catastrophic disruptions, from jeopardizing passenger safety to compromising personal data. Protecting against threats involves a multi-layered approach, incorporating robust encryption methods, secure communication protocols, and regular software updates to patch vulnerabilities. Additionally, collaboration between vehicle manufacturers, cybersecurity experts, and regulatory bodies is essential to develop stringent standards and best practices. This collaborative effort aims to ensure that autonomous vehicles not only operate efficiently but also maintain the highest levels of security against ever-evolving cyber threats.

Protecting Against Threats

The landscape of autonomous vehicles represents a paradigm shift in transportation, offering unparalleled convenience, efficiency, and safety. However, as with any advancing technology, it also ushers in new and complex security challenges. Ensuring that self-driving cars are protected against cyber threats is not just a technical necessity but a foundational requirement for the technology's widespread adoption.

One of the foremost threats to autonomous vehicles lies in the potential for unauthorized access to the vehicle's systems. Hackers could exploit vulnerabilities in the car's software, hardware, or communication channels. This could range from relatively benign intrusions, such as unauthorized access to vehicle telemetry, to more malicious attacks, such as commandeering the vehicle's controls. Cybersecurity in autonomous vehicles must, therefore, prioritize robust authentication and encryption protocols to prevent any unauthorized access and ensure data integrity.

Another critical threat involves the potential for Denial of Service (DoS) attacks. In a DoS attack, the perpetrator seeks to render a network resource unavailable to its intended users, typically by overwhelming the system with a flood of malicious requests. For autonomous vehicles, a successful DoS attack could disable critical systems, potentially causing accidents or stranding passengers. Implementing resilient network infrastructure and developing sophisticated intrusion detection systems is essential to mitigate this risk.

Moreover, the integrated nature of Vehicle-to-Everything (V2X) communication poses its own unique challenges. V2X communication allows vehicles to interact with various entities, including other vehicles, road infrastructure, and cloud services. While this connectivity significantly enhances the vehicle's capability to make real-time decisions, it also introduces vulnerabilities that could be exploited by cyber criminals. Secure communication protocols that use end-to-end encryption are crucial to ensuring that data exchanged within the V2X ecosystem is safeguarded against interception and manipulation.

Protecting the software supply chain is also paramount. Autonomous vehicles rely on a multitude of software components often sourced from third-party vendors. Any compromise within these supply chains can introduce vulnerabilities into the vehicle's core systems. Regular auditing of software components, coupled with stringent code review processes, helps in identifying and mitigating potential weaknesses before they can be exploited.

In addition to external threats, insider threats should not be overlooked. Employees with access to proprietary systems may inadvertently or maliciously introduce vulnerabilities. Establishing rigorous access controls and routinely monitoring employee activity within the system can help mitigate such risks. Furthermore, investing in comprehensive employee training programs can educate staff on the latest cybersecurity practices and foster a culture of security awareness.

Another component of securing autonomous vehicles is ensuring robust data privacy measures. With autonomous vehicles continually collecting vast amounts of data—including not just operational data but also personal data related to passengers—securing this information against unauthorized access is critical. Implementing stringent data protection regulations in line with global standards, alongside robust data anonymization techniques, can help protect personal information and maintain user trust.

Additionally, regular security testing and updates are fundamental to maintaining the integrity of an autonomous vehicle's systems. Periodic vulnerability assessments and penetration testing (pen-testing) can help identify and rectify security flaws. Moreover, maintaining an efficient update mechanism ensures that any newly discovered threats or vulnerabilities can be promptly patched, thereby enhancing the vehicle's resilience against evolving cyber threats.

Collaboration with regulatory bodies is another key aspect of fortifying cybersecurity in autonomous vehicles. Establishing industry-wide standards and compliance requirements can create a unified approach to cybersecurity, driving all stakeholders towards common goals and helping to ensure that all vehicles meet a baseline level of security. Moreover, fostering public-private partnerships can facilitate the sharing of threat intelligence, leading to more effective responses to emerging threats.

Given the global nature of the automotive industry, international collaboration can also play a vital role. Harmonizing cybersecurity regulations and standards across different jurisdictions can help create a seamless protection framework that transcends national borders, making it more difficult for cyber criminals to exploit geographical loopholes.

The integration of artificial intelligence (AI) within cybersecurity solutions is another promising avenue for enhancing the protection of autonomous vehicles. AI-driven security measures can more effectively monitor network traffic, analyze patterns, and detect anomalies indicative of potential threats. Leveraging machine learning algorithms allows for automated and adaptive responses to cyber threats, further bolstering the defensive infrastructure of autonomous vehicles.

To remain ahead of potential threats, automotive companies must adopt a proactive rather than a reactive approach to cybersecurity. This involves not only addressing existing vulnerabilities but also anticipating future risks. Continuous research and development in cybersecurity technologies, along with active participation in cybersecurity forums and industry consortia, can help companies stay abreast of the latest threat landscapes and effectively adapt their security strategies.

Finally, consumer education and trust are integral to the success of autonomous vehicles. Transparent communication about the cybersecurity measures in place, alongside guidance on user best practices for maintaining vehicle security, can help build public confidence. Ensuring that cybersecurity considerations are woven into the fabric of autonomous vehicle design from the outset will go a long way in ensuring a secure, safe, and reliable future for self-driving cars.

In conclusion, while the advent of autonomous vehicles presents a game-changing leap in transportation, it simultaneously introduces a complex array of cybersecurity challenges. Through a comprehensive, multi-faceted approach that includes robust technical defenses, stringent regulatory compliance, continuous innovation, and public engagement, the industry can effectively safeguard against these threats. Ensuring the cybersecurity of autonomous vehicles is not just a technical hurdle but a pivotal component of their successful integration into society.

Secure Communication Protocols

The intricacy of secure communication protocols in autonomous vehicles can't be overstated. As the brains behind self-driving cars, these protocols ensure the safe and reliable exchange of data between vehicles, infrastructure, and other connected systems. Without robust communication channels, the very idea of autonomous driving collapses under potential threats and malfunctions.

Let's break down what secure communication entails. In essence, it's about establishing methods to transmit data without unauthorized access or alteration. For autonomous vehicles, this security layer is non-negotiable. They rely on vehicular networks to make split-second decisions, and any vulnerability could lead to catastrophic outcomes, both in terms of safety and trust.

First off, we need encryption. Encryption transforms readable data into a coded format, decipherable only by those possessing the correct key. For autonomous vehicles, encryption ensures that any data transmitted—from GPS coordinates to sensor readings—is protected from tampering. Think of it as a lock and key system; even if someone intercepts the data, it's meaningless without the key.

Public Key Infrastructure (PKI) plays a crucial role here. PKI uses a pair of keys: a public key and a private key. The public key encrypts the data, while the private key decrypts it. This system ensures that only authorized entities can access sensitive data. PKI also incorporates digital certificates to verify identities, adding an extra layer of trust and integrity.

In addition to encryption, secure communication protocols demand authentication. Authentication verifies whether the entities involved in the data exchange are who they claim to be. Secure protocols use methods like digital signatures and certificates to confirm the identity of the communicating parties. This prevents spoofing attacks where a malicious entity poses as a legitimate one.

Beyond encryption and authentication, integrity is a vital component. Data integrity ensures that the information remains unchanged during transit. Hash functions are commonly used to achieve this. They create a unique fingerprint for each data set, known as a hash value. When the data reaches its destination, the receiving system calculates its hash value and compares it to the original. A mismatch indicates tampering, prompting immediate action.

The next phase involves intrusion detection systems (IDS). IDS continually monitor the network for any unusual activity or potential breaches. If an anomaly is detected, the system alerts administrators to take immediate action. Time is of the essence here, as even a small lapse can have serious repercussions.

One can't ignore the importance of secure protocols in Vehicle-to-Everything (V2X) communication. V2X enables vehicles to communicate with each other (V2V) and with infrastructure (V2I). Messages exchanged in this network are crucial—think traffic updates, hazard notifications, or even emergency vehicle alerts. The use of secure protocols ensures that these messages are authentic and reach their intended recipients without delay.

Moreover, the hardware itself must be secure. Secure boot mechanisms help in this regard. They prevent malicious software from loading during the vehicle's startup process. Essentially, the system checks each piece of code against a known good set, refusing to boot if discrepancies appear. It's like a bouncer not letting troublemakers into a secure club.

Nevertheless, with increased connectivity comes increased risk. Therefore, cybersecurity measures need constant updates. Over-the-air (OTA) updates are essential for rolling out security patches and software enhancements without the need for physical recalls. OTA updates ensure that all vehicles in the network stay up-to-date with the latest protection mechanisms.

Lastly, consider the role of artificial intelligence in enhancing secure communications. AI algorithms can predict potential threats by analyzing massive data streams in real-time. Machine learning models can identify patterns of normalcy and flag irregular activities that deviate from these patterns. This proactive approach catches potential issues before they escalate into full-blown attacks.

It's also worth mentioning that collaboration plays a critical role. Automakers, cybersecurity firms, regulators, and researchers must work together to create and maintain secure communication protocols. Standardized practices and open sharing of information help bolster the entire ecosystem against threats. This collaborative effort ensures that no single point of failure can compromise the system.

In summary, secure communication protocols are the linchpin of the autonomous vehicle infrastructure. They protect against unauthorized access, ensure data integrity, and keep the system resilient against cyber-attacks. Encryption, authentication, PKI, IDS, and secure hardware collectively form the defensive shield that makes autonomous driving safe. By continually evolving these protocols, we can pave the way for a future where the roads are dominated by safe, reliable, and secure autonomous vehicles.

Chapter 15:
AI in Fleet Management

A I is revolutionizing fleet management by significantly enhancing operational efficiency and reducing costs. Through sophisticated optimization techniques, companies can now manage their fleets with unprecedented precision. AI algorithms analyze vast amounts of data to optimize routes, schedules, and loads, ensuring each vehicle operates at peak efficiency. Predictive maintenance powered by AI keeps vehicles in top condition, reducing downtime and preventing costly repairs by forecasting issues before they occur. This harmonious blend of AI technologies transforms fleet management into a proactive, streamlined process, aligning perfectly with the ever-evolving demands of modern transportation systems.

Optimization Techniques

AI in fleet management can be a game-changer in transportation, offering a broad array of optimization techniques that ensure efficient and cost-effective operations. These methods aim to address crucial aspects such as route optimization, fuel efficiency, driver performance, and vehicle health. Each of these areas not only enhances the overall productivity of fleets but also minimizes operational costs.

Route optimization is one of the primary sectors where AI makes a significant impact. By leveraging advanced algorithms and real-time data, AI-driven systems can determine the most efficient paths for vehicles to take. This not only reduces fuel consumption but also shortens delivery times and improves customer satisfaction. Factors such as traffic conditions, weather, and roadwork are continuously analyzed to provide the most efficient route suggestions.

Another critical optimization technique focuses on fuel efficiency. AI can monitor driving patterns and suggest improvements for better fuel economy. For instance, it can recommend optimal speeds, acceleration patterns, and even idling times to save fuel. By implementing such data-driven strategies, companies can significantly reduce their carbon footprints and operational costs.

Driver performance is another area where AI technologies play a pivotal role. With real-time monitoring and analytics, fleet managers can track key performance indicators such as adherence to speed limits, harsh braking, and acceleration events. AI systems can provide instant feedback to drivers, helping them adopt safer and more efficient driving habits. Additionally, this data can be used for training programs tailored to improve driver behavior and overall fleet safety.

Predictive maintenance is an emerging trend powered by AI that can preemptively address vehicle health issues before they escalate into significant problems. By analyzing data from various sensors within the vehicle, AI systems can predict when a component is likely to fail and schedule maintenance accordingly. This not only reduces downtime and maintenance costs but also extends the lifecycle of the fleet.

In terms of vehicle health, AI-driven diagnostics tools can constantly monitor and evaluate the state of each vehicle in the fleet. Early detection of issues such as tire wear, brake deterioration, and engine malfunctions means that proactive measures can be taken, thereby avoiding costly repairs and ensuring that the fleet runs smoothly. This real-time diagnostics capability also helps in dynamically adjusting maintenance schedules, reducing unnecessary servicing and downtime.

Beyond individual vehicle optimization, AI also facilitates better fleet-wide management strategies. This includes load optimization, where AI can determine the best way to distribute cargo across the fleet to ensure balanced loads. It can also recommend vehicle assignments based on criteria like fuel efficiency, current vehicle health, and driver availability, thus further optimizing the overall fleet performance.

One of the newer methods being explored is the integration of AI with Vehicle-to-Everything (V2X) communication. This allows vehicles to communicate with each other and with infrastructure to optimize routes and driving patterns even further. For instance, a fleet could adjust speeds and routes in real-time based on data from traffic signals, road sensors, and other vehicles, leading to more streamlined and efficient operations.

In terms of environmental impact, reducing emissions is a key focus area. AI-driven optimization techniques can help identify the most fuel-efficient routes and operational strategies. By forecasting traffic congestion and weather conditions, AI systems can recommend times and routes that consume less fuel and produce fewer emissions. This aligns with broader sustainability goals and helps companies adhere to environmental regulations.

Artificial intelligence isn't just making fleets smarter; it's making them greener. Predictive analytics allows for better energy usage, from managing fuel consumption to integrating electric vehicle technologies. AI can analyze battery performance, optimal routes for conserving energy, and even identify suitable charging points along the route.

Furthermore, AI in fleet management extends to supply chain interactions. By integrating AI systems with supply chain logistics, fleets can ensure that goods are transported in the most efficient manner. These systems can predict demand, adjust supply routes, and even reroute shipments in real-time to meet changing circumstances, thereby optimizing overall supply chain performance.

Fleet management also benefits from AI when handling complex scheduling tasks. AI algorithms can efficiently manage schedules by considering various constraints, such as driver availability, hours of service regulations, and customer delivery windows. This helps in minimizing delays while ensuring compliance with regulatory standards. Advanced scheduling tools can also offer contingency plans in case of unforeseen events, making fleet operations more resilient.

In essence, optimization through AI transforms fleet management by offering a holistic approach that encompasses everything from route planning to maintenance and real-time decision-making. By continuously learning and adapting to new data, AI systems provide an evolving optimization framework that can keep up with the dynamic nature of transportation logistics. These optimization techniques don't just make fleets faster and cheaper to run; they also make them safer and more sustainable, ensuring that businesses can meet both their operational and environmental goals.

Predictive Maintenance

Predictive maintenance is a game-changer in the realm of AI in fleet management. It hinges on the idea of foreseeing and addressing mechanical failures before they occur, thus drastically reducing downtime and maintenance costs. Predictive maintenance employs a blend of sensors, big data, and machine learning algorithms to constantly monitor the condition and performance of vehicles. By analyzing this continual inflow of data, fleet managers can pinpoint the exact moment when a component might fail, allowing for timely intervention and repairs.

Potential issues like engine malfunctions, brake wear, or tire failures can be detected early. For instance, sensors embedded in critical vehicle components continuously transmit data points on temperature, vibration, noise, and other parameters. AI algorithms then process these massive datasets in real time, identifying patterns and anomalies that signify impending problems. This proactive approach stands in stark contrast to traditional maintenance, which typically operates on fixed schedules or after a failure occurs.

One major advantage of predictive maintenance is cost efficiency. Repairing a vehicle before a failure is not only cheaper but also less time-consuming. Fleet managers can schedule maintenance in a way that minimizes disruptions, ensuring that vehicles are maintained during off-peak hours or when a replacement vehicle is available. This strategy keeps the fleet operational and significantly improves customer satisfaction by reducing delays and service interruptions.

Furthermore, predictive maintenance enhances safety. By addressing mechanical issues before they escalate, the risk of accidents due to equipment failure can be substantially lowered. This is particularly critical for fleets involved in long-haul transportation or public transit, where vehicle reliability directly impacts passenger safety and trust. Predictive algorithms also aid in optimizing the lifespan of vehicle components, ensuring that they are replaced precisely when needed rather than too early or too late.

Another less obvious benefit is the impact on inventory management. Knowing in advance which parts are likely to fail allows fleet operators to maintain an optimal stock of spare parts. This eliminates the need for emergency orders and reduces the storage costs associated with keeping an extensive inventory of various parts. The improved efficiency not only reduces costs but also leads to better utilization of warehouse space.

Cutting-edge machine learning models lie at the heart of predictive maintenance. These models can learn from historical data and continuously adapt to new data inputs, refining their accuracy over time. Consider a model trained on data from thousands of vehicles over several years – it becomes incredibly adept at predicting failures based on the smallest deviations from normal operation. The more data the system ingests, the more precise and reliable it becomes.

Implementing predictive maintenance does come with its challenges. The initial setup involves significant investment in sensors and data infrastructure. Additionally, there is a need for high-quality data to train the machine learning models. Poor data quality or incomplete datasets can lead to inaccurate predictions, undermining the system's effectiveness. Hence, ongoing data management, cleansing, and governance are crucial to maintain the integrity and efficacy of these predictive models.

Another challenge lies in integrating predictive maintenance systems with existing fleet management software. Many fleet operators use legacy systems that may not be compatible with modern AI-driven tools. This necessitates either upgrading existing software or developing custom integrations, which can be both time-consuming and costly. However, the long-term benefits typically outweigh these initial hurdles.

As AI continues to evolve, so too will the capabilities of predictive maintenance systems. Advances in edge computing, for instance, enable real-time data processing directly on the vehicle, reducing latency and enhancing the system's responsiveness. Similarly, developments in Internet of Things (IoT) technologies are leading to more advanced and affordable sensors, broadening the scope and scale of data collection.

The impact of predictive maintenance extends beyond individual fleet operators. Entire industries could be transformed through the widespread adoption of these practices. For example, logistics companies can ensure timely delivery of goods, enhancing the reliability of supply chains. Public transit systems adopting predictive maintenance can provide more consistent and safer service, improving the overall quality of urban mobility.

The societal benefits are not to be understated either. Reduced vehicle downtime and enhanced efficiency contribute to lower emissions and a smaller carbon footprint. As fleets operate more smoothly and predictably, traffic congestion can be alleviated, contributing to a more streamlined and less stressful driving environment for everyone on the road.

Moreover, predictive maintenance fosters a culture of continuous improvement. Instead of reacting to breakdowns, fleet managers can focus on optimizing vehicle performance, lifecycle management, and overall operational excellence. This shift from reactive to proactive maintenance has the potential to redefine the industry standards, setting new benchmarks for efficiency and reliability.

Ultimately, the integration of predictive maintenance within AI-driven fleet management encapsulates the essence of what technology promises for the future: intelligent, efficient, and safer systems. It provides a glimpse into how AI can transform traditional practices, offering tangible benefits that resonate across multiple dimensions – from operational efficiency and cost savings to safety and environmental sustainability.

Chapter 16:
Autonomous Driving in Rural vs. Urban Settings

As autonomous vehicles navigate their way into mainstream transportation, the contrast between rural and urban environments presents unique challenges and solutions that are essential to understand. In urban settings, the complexity of dense traffic, frequent stops, and numerous pedestrians requires advanced sensor integration and split-second decision-making by AI systems. On the other hand, rural areas, with their sparse infrastructure and longer distances, demand robust GPS accuracy and reliable communication in often low-connectivity zones. While urban landscapes offer the advantage of more accessible data for AI learning, rural environments push the boundaries of autonomous driving technology with fewer landmarks and unpredictable road conditions. Balancing these contrasting needs will be pivotal for developers aiming to ensure that self-driving technology can provide safe and efficient transportation across diverse geographies.

Infrastructure Differences

When discussing autonomous driving vehicles, the contrast between rural and urban settings reveals significant infrastructure differences that impact the deployment and efficiency of these advanced systems. Urban areas typically offer a more predictable environment characterized by well-maintained roads, uniform traffic signals, and numerous data points for AI systems to analyze. However, the complexity and density of these areas introduce their own set of challenges for autonomous vehicles.

In urban settings, the road infrastructure is well-developed with clearly marked lanes, extensive sensor networks, and smart traffic signals. These features provide a conducive environment for the deployment of autonomous vehicles. The existence of data-rich environments helps autonomous systems to process real-time information effectively, thanks to a continuous flow of data from various sources like cameras, LIDARs, and V2X communication systems. Additionally, urban zones often have high connectivity infrastructure, enabling faster data exchange and real-time updates critical for navigating crowded streets and avoiding obstacles.

Urban environments, despite their advantages in infrastructure, present intricate challenges due to the density and unpredictability of pedestrian and vehicular traffic. The presence of numerous crosswalks, cyclists, and unpredictable pedestrian behaviors necessitate highly sophisticated AI systems capable of making split-second decisions. The traffic congestion in cities can also be erratic, with peak hours bringing in gridlocks that require seamless coordination among self-driving cars to optimize traffic flow.

Rural settings, on the other hand, introduce a different set of challenges and considerations for autonomous driving technology. Here, the infrastructure tends to be less developed; roads may lack clear markings, consistent signage, or proper maintenance. The absence of detailed mapping and irregularities in road conditions can test the robustness of autonomous driving algorithms. Sensor systems must be more adaptive, with AI compensating for the lack of established infrastructure through advanced path planning and obstacle detection.

Despite the apparent challenges, rural areas offer certain advantages for autonomous vehicles. The lower density of both pedestrian and vehicular traffic reduces the risk of complex interactions and potential accidents. The expansive and often empty roads provide an excellent testing ground for the capabilities of self-driving technologies. In these settings, the AI systems can benefit from simpler scenarios where unexpected elements like jaywalking pedestrians or dense traffic are less common.

Moreover, rural environments might not yet have wide-scale implementation of V2X communication systems, necessitating autonomous vehicles to rely more heavily on onboard sensors and machine learning algorithms to understand and navigate the environment safely. With fewer traffic signals, stop signs, and roundabouts, the system must be adept at interpreting natural cues and adapting to the road's conditions.

Notably, the deployment strategies for autonomous vehicles in rural settings might differ significantly from those in urban environments. For instance, rural areas could rely more on service hubs located in small towns, where vehicles could return for updates and maintenance. This decentralized approach contrasts sharply with the centralized model commonly seen in urban areas, where vehicles can avail themselves of services and updates almost anywhere within the city.

Focus shifts to road conditions and materials employed in various settings. Urban areas might boast roads made from advanced materials designed to withstand heavy traffic and maintenance activities. Consequently, autonomous vehicles here can expect a predictable environment with fewer surprises. Conversely, rural roads may vary from tarmac to gravel and dirt, each presenting unique hurdles in terms of traction and control. The autonomous systems, therefore, need to be more versatile and resilient to a wide array of conditions.

The contrast extends to communication infrastructure, as well. Urban settings often enjoy the luxury of high-speed internet and robust cellular networks, facilitating real-time vehicular communication and navigation updates. These digital advantages are less reliable in rural regions, where network coverage can be patchy at best. Autonomous vehicles operating in such environments must be capable of local data processing and decision-making without relying on constant connectivity.

From a regulatory perspective, urban areas are more likely to have stringent rules and regulations specifically crafted to accommodate the complexities of autonomous vehicles. The legislative framework might include dedicated lanes for autonomous vehicles, specific guidelines for interaction with human-driven vehicles, and extensive data security protocols. On the other hand, rural regions may lag behind in this aspect, necessitating custom adaptations of existing traffic laws to make autonomous driving feasible.

The role of infrastructure investment can't be overstated. Urban planners and governments might prioritize investments in smart traffic management systems, V2X infrastructure, and road quality improvements to support the burgeoning growth of autonomous vehicles. In contrast, rural areas might struggle with limited budgets, resulting in slower adaptation and lesser incentives for infrastructure enhancement. Overcoming these disparities requires collaborative efforts between public agencies, private companies, and local communities to ensure that the benefits of autonomous driving can be realized universally.

Interestingly, the differences in infrastructure also spell out varying economic implications. Urban settings with advanced, well-funded infrastructures might serve as initial profit centers for companies deploying autonomous taxis, ride-sharing services, or delivery vehicles. The high density of potential users and robust infrastructure support creates an economically viable environment. In rural settings, a more significant challenge lies in achieving profitability due to the lower population density and fewer potential users.

Connecting rural communities with autonomous vehicle networks can provide profound societal benefits, such as reduced transportation deserts and improved accessibility. However, this requires a paradigm shift in how infrastructure is planned and implemented. It necessitates an increase in localized charging stations for electric autonomous fleets and improved road maintenance activities to cater to these advanced vehicles.

It's clear that the infrastructure differences between rural and urban settings pose unique challenges and opportunities for the deployment of autonomous vehicles. To realize the full potential of autonomous driving technologies, adaptations and strategic investments must be tailor-fit to each environment's particular needs. Only through such thoughtful consideration can we ensure that both urban and rural landscapes are prepared for the transformative changes autonomous vehicles bring to the transportation sector.

Unique Challenges and Solutions

Autonomous driving presents unique challenges when considering rural and urban settings, each with distinct environmental conditions and operational requirements. Understanding and addressing these challenges are crucial for the successful deployment of self-driving vehicles across diverse landscapes.

In urban settings, the primary challenge lies in the high density of both traffic and pedestrians. Urban areas are characterized by complex road networks. Multiple lanes, numerous intersections, and frequent changes in traffic signals demand quick decision-making from autonomous systems. The unpredictability of human behavior, such as jaywalking or sudden lane changes, adds another layer of complexity.

Advanced machine learning algorithms and extensive sensor arrays are essential to navigating these environments safely.

On the flip side, rural areas present different but equally daunting obstacles. The lower density of human activity means fewer vehicles and pedestrians, but the absence of well-defined infrastructure can be problematic. Rural roads often lack the clear lane markings and consistent signage found in cities. Moreover, these areas are more likely to have irregular terrains, including gravel roads and unpaved paths, which can pose significant challenges for sensor accuracy and vehicle stability.

One major solution to urban challenges is the development and deployment of high-definition maps coupled with real-time data feeds. High-definition mapping technology provides autonomous vehicles with detailed information about road layouts, traffic signals, and pedestrian zones. This allows for more precise navigation. Real-time data feeds, collected from various sources such as cameras and radar systems, help the vehicle make instantaneous decisions, improving its ability to handle unexpected scenarios.

In rural settings, where the infrastructure is often minimal, the focus shifts to enhancing the robustness and adaptability of sensor technology. Lidar, radar, and advanced camera systems must be fine-tuned to detect and interpret irregularities in the terrain. Additionally, leveraging satellite-based navigation systems that offer high precision can help in areas where road markings are sparse or non-existent.

Weather conditions are another factor where rural and urban environments differ significantly. Urban areas, with their dense concentration of buildings, often experience microclimates that can affect visibility and road conditions. Automated systems need to manage scenarios involving heavy rain, fog, or snow, all of which can obscure sensors and challenge machine vision systems. Rural areas, while generally offering clearer skies, can have extreme weather events that affect terrains, such as flooding or snow accumulation on roads, making travel hazardous.

Machine learning models, particularly deep learning techniques, are being trained to predict and respond to a variety of weather conditions. In urban areas, this involves integrating data from multiple local sensors to create a comprehensive environmental model. Weather forecasting can be integrated into navigational algorithms to help vehicles prepare for upcoming conditions, enhancing their safety and reliability.

Another urban challenge is the need for robust cybersecurity measures to prevent vulnerabilities. City environments, with their high connectivity and extensive use of IoT devices, are susceptible to cyber-attacks. Autonomous vehicles must be equipped with secure communication protocols to safeguard against these threats. Conversely, rural areas might have less connectivity infrastructure, leading to challenges in maintaining constant data flow for vehicle-to-everything (V2X) communication. This requires building more resilient systems capable of operating independently of continuous connectivity.

Urban settings also demand more efficient energy consumption plans due to frequent stop-and-go traffic. Electric vehicles, which are often used in autonomous fleets, need optimized battery management systems to ensure they can handle the urban gridlock without frequent recharging. This can be addressed by implementing predictive algorithms that optimize energy usage based on real-time traffic data and route planning.

In rural areas, the challenge may be the lack of charging infrastructure. Implementing widespread autonomous driving in these regions means addressing the scarcity of charging stations. One solution involves developing longer-lasting batteries and more efficient energy consumption systems. Solar-powered charging stations could be another viable solution, especially in remote areas with ample sunlight.

Additionally, the societal and economic implications of autonomous driving differ between these settings. Urban areas might see significant shifts in labor markets, particularly in jobs related to driving and transportation services. There's a potential for job displacement that needs to be managed through retraining programs and policy interventions. Rural areas, though less affected by job displacement in transportation, may experience economic changes due to improved accessibility. Autonomous vehicles could open up rural regions to greater economic opportunities by facilitating better connectivity.

Public acceptance also varies. Urban residents, who are more accustomed to technological advancements and infrastructure changes, may be quicker to adopt autonomous vehicles. In contrast, rural communities, which often have a more conservative approach to new technology, might require more extensive education and engagement efforts to build trust in autonomous systems.

Both environments can benefit from community-led initiatives to foster acceptance and understanding. In urban areas, pilot projects in specific districts can demonstrate the benefits and safety of autonomous driving. For rural areas, workshops and community events can provide hands-on experience with autonomous technology, helping to alleviate concerns and build trust.

The integration of autonomous vehicles into public transportation systems presents another unique challenge and opportunity. Urban areas with extensive public transport networks can see autonomous vehicles as a supplement to existing services, offering first-mile and last-mile connectivity. This can alleviate congestion and improve the efficiency of the transport ecosystem. Conversely, in rural areas, where public transportation is often limited, autonomous vehicles can provide essential services, improving access to healthcare, education, and employment.

Infrastructure adaptation is essential for both settings. Urban areas need to upgrade their traffic management systems to interact seamlessly with autonomous vehicles. This includes smart traffic lights and dedicated lanes for self-driving cars. Rural areas, on the other hand, may require upgrades to basic infrastructure, such as road maintenance and reliable internet connectivity, to support the data needs of autonomous vehicles.

Lastly, the role of government policy is indispensable in addressing these challenges. Regulatory frameworks must be adaptable to the differing needs of urban and rural deployments. Urban policies may focus on managing high traffic volumes and ensuring cybersecurity, while rural policies might emphasize infrastructure development and connectivity improvements.

In summary, while both rural and urban settings face unique challenges in adopting autonomous vehicles, the solutions lie in tailored technological advancements, innovative infrastructure improvements, and targeted policy frameworks. By addressing these specific needs, the benefits of autonomous driving can be realized across diverse environments, paving the way for a safer, more efficient, and inclusive transportation future.

Chapter 17:
The Impact on Personal Car Ownership

The advent of autonomous driving technology is poised to reshape personal car ownership in profound ways. Consumers may begin to question the need for owning a vehicle when autonomous ride-sharing and subscription services offer convenient, cost-effective alternatives. The traditional model of purchasing and maintaining a car could give way to a landscape dominated by shared, on-demand access to autonomous fleets. This shift would not only reduce the number of cars on the road but also transform urban planning, as parking spaces become less critical and cities can repurpose land for other uses. Furthermore, the environmental benefits, paired with the economic incentives for both consumers and manufacturers, signal a future where personal car ownership is no longer the norm but a premium choice. The potential for such significant changes calls for a deep understanding of consumer behavior, adjustments by auto companies, and adaptive regulatory frameworks to address the emerging trends and challenges in a rapidly evolving automotive landscape.

Shifts in Consumer Behavior

The wave of autonomous driving technology has undeniably transformed consumer behavior in significant ways. There's been a noticeable shift in attitudes toward personal car ownership, with many reconsidering the necessity of owning a car. For urban dwellers, ride-sharing platforms, which utilize autonomous vehicles, offer a compelling alternative to the traditional model of one person, one car. This shift is partly driven by the convenience and cost-effectiveness that these services promise, making the economics of car ownership appear less attractive.

In addition to convenience, environmental consciousness is playing a crucial role. Many consumers are becoming increasingly aware of their carbon footprint. Autonomous electric vehicles are often perceived as a greener alternative to conventional cars, leading environmentally conscious consumers to favor shared autonomous vehicles over owning a personal car. The rising awareness of environmental issues and the appeal of reducing individual carbon footprints cannot be underestimated in this shift.

Moreover, the integration of advanced AI and seamless connectivity has also reshaped consumer expectations. People now expect their commutes and travel experiences to be enriched with real-time updates, personalized settings, and high levels of efficiency and safety. Autonomous vehicles equipped with sophisticated AI cater to these evolving preferences by offering an optimized and tailored travel experience, reducing the need for individual car ownership.

One cannot overlook the demographic changes contributing to this behavioral shift. Younger generations, particularly millennials and Gen Z, prioritize experiences over possessions. The notion of car ownership, a symbol of freedom and status for previous generations, doesn't hold the same appeal for them. They prefer the flexibility of accessing on-demand mobility services rather than dealing with the responsibilities and costs associated with owning a car. This reflects a broader trend of valuing access over ownership in various aspects of life.

Furthermore, the rise of remote work and the gig economy has influenced how people think about transportation needs. With many employers adopting flexible work arrangements, the daily commute is no longer a necessity for many workers. This reduced need for daily travel diminishes the perceived necessity of owning a car, particularly in urban areas with robust public transportation and ride-sharing options.

However, this shift is not uniform across all demographics or geographies. In rural areas, where public transportation options are limited, the need for personal vehicles remains high. Despite the advancements in autonomous technology, there are still unique challenges in reaching remote and less densely populated regions. These differences highlight the complexity of consumer behavior and the varying factors that influence the decision to own a car.

Safety concerns also play a part in this decision-making process. As autonomous vehicles continue to prove their reliability through extensive testing and real-world usage, trust in these technologies grows. However, a segment of consumers remains skeptical, tethered to the perceived safety and control offered by traditional, manually-operated vehicles. The gradual increase in public confidence in autonomous systems will be critical in further shifting consumer preferences.

Cost considerations are another major driver. The total cost of vehicle ownership – including purchase price, maintenance, insurance, and fuel – is substantial. Autonomous ride-sharing services can lower these costs significantly for consumers, making it more economical to opt for on-demand transportation solutions. This economic advantage is a powerful incentive for consumers to transition away from personal ownership.

Insurance models, too, are adapting to these shifts. Traditional car insurance is being reevaluated to accommodate the unique risk profiles of autonomous vehicles. Usage-based insurance models, which align more closely with the operational patterns of ride-sharing and autonomous vehicle usage, provide further financial motivation for consumers to reconsider car ownership. These evolving insurance paradigms influence consumer decisions by possibly lowering the costs associated with driving.

Yet, it's important to note the sentimental value attached to car ownership. For many, cars are not just a mode of transportation but a point of personal pride and emotional attachment. Vehicles often embody personal identity and are linked to significant life events and memories. Overcoming this sentimental value is another hurdle for the widespread adoption of shared autonomous vehicle solutions.

In summary, shifts in consumer behavior are influenced by a complex interplay of convenience, environmental concerns, technological advancements, economic factors, demographic changes, safety perception, and emotional attachment. As autonomous driving technology continues to develop and gain acceptance, these shifts are likely to become more pronounced, leading to a future where personal car ownership is redefined. The evolving landscape promises a transportation paradigm that prioritizes flexibility, efficiency, and sustainability, significantly altering how consumers approach their mobility needs.

Future Sales and Ownership Models

The landscape of vehicle ownership is set to undergo a significant metamorphosis in the coming years, predominantly driven by the integration of autonomous driving technologies. As we advance, traditional personal car ownership, characterized by individual car purchases and private maintenance, faces considerable disruption. Instead, we are likely to see the emergence of more fluid and dynamic ownership models that reflect the versatile nature of autonomous vehicles.

A compelling future sales model is the concept of subscription-based car services. Much like the streaming services revolution in digital media, consumers could subscribe to a car service that provides them access to a fleet of autonomous vehicles. This model holds distinct advantages, such as reducing the financial burden of car loans, insurance, and maintenance costs. It allows consumers to choose from a variety of vehicles depending on their immediate needs—whether it's a compact car for city commuting or an SUV for a weekend getaway.

In addition to subscription services, the concept of car-sharing is poised for exponential growth. Companies like Zipcar and Turo have already paved the way, but autonomous vehicles can elevate this to a whole new level. Self-driving cars could be booked via an app, arrive at your location without a human driver, and then return themselves to a designated area once the journey is complete. This increased convenience and efficiency could reduce the need for owning a personal vehicle, especially in urban areas where parking and congestion are constant issues.

Another transformative factor is the potential for autonomous vehicle fleets to be owned and operated by corporations rather than individuals. Imagine a future where companies like Uber or Lyft own vast fleets of autonomous vehicles, providing rides to consumers on demand. These fleets could operate around the clock, optimizing routes and reducing downtime, ultimately lowering ride costs for consumers. For corporations, this model provides the benefit of scale and centralized maintenance, further enhancing operational efficiency.

Peer-to-peer (P2P) car sharing is another model that could see substantial uptake. In this scenario, individuals who do own autonomous vehicles can rent them out to others when not in use. P2P platforms would facilitate these transactions, allowing car owners to earn income from their autonomous cars while they are idle. This model not only maximizes asset utilization but also encourages a community-oriented approach to vehicle usage.

The potential reduction in personal car ownership has significant implications for auto manufacturers. Traditional sales numbers may decline, prompting these companies to innovate not just in technology but in business models as well. Some might pivot to become mobility service providers, offering complete packages that include not just the vehicle, but also maintenance, insurance, and other associated services. This shift requires car manufacturers to rethink their distribution networks, customer engagement strategies, and even their production processes.

Furthermore, the resale market for cars could be fundamentally altered. Autonomous vehicles might be designed with modular components that can be easily updated, extending their usable lifespan and reducing the frequency of new car purchases. These vehicles could be 're-skinned' with new interiors and exteriors, or updated with the latest AI software, keeping them current and functional for longer periods. Consequently, the notion of a car diminishing rapidly in value after a few years could become obsolete.

Financing and leasing models will also adapt to the new landscape. Financial institutions and leasing companies might offer more flexible terms to attract customers who are hesitant to commit to long-term ownership. Pay-per-use models and micro-leases could become standard, allowing consumers to lease a vehicle for days, hours, or even minutes, depending on their needs. These models make owning a vehicle more financially accessible to a broader audience, thus democratizing access to advanced autonomous technology.

The changing landscape will also influence insurance models. Traditional auto insurance is primarily predicated on driver behavior and ownership risk. As we move toward autonomous vehicles and shared ownership, insurance companies will need to adapt by focusing more on the technological reliability and operational uptime of these cars. Policies might shift from individual plans to fleet-based coverage or usage-based insurance where premiums are calculated based on when and how often a vehicle is used.

One fascinating speculation is the possibility of dynamic pricing in car sales and rentals. Autonomous vehicles equipped with advanced analytics could monitor usage patterns, current demand, and even seasonal trends to dynamically adjust prices in real-time. Such a system would benefit consumers through competitive pricing while allowing service providers to maximize revenue.

Further contributing to the shift in ownership models is the integration of smart city infrastructure. With better V2X communication and real-time data analytics, cities could create more efficient transportation networks that capitalize on the capabilities of autonomous vehicles. Municipalities might own and operate fleets of autonomous vehicles for public use, blending them seamlessly into public transportation networks and reducing the need for private vehicle ownership even further.

Additionally, as environmental concerns continue to grow, there is a push towards more sustainable and eco-friendly transportation solutions. Autonomous vehicles, particularly electric ones, align well with this objective. Governments and regulatory bodies may incentivize shared and subscription-based models with tax breaks, subsidies, or priority lanes, further accelerating their adoption and pushing consumers away from owning personal cars.

A paradigm shift in personal car ownership signifies not just an economic transition but a cultural one as well. Societies that have long valued the independence and status symbol afforded by car ownership may gradually shift towards valuing convenience, accessibility, and sustainability more. This cultural evolution could stimulate a more community-focused approach to transportation, promoting shared responsibility and efficient resource utilization.

As autonomous vehicles become ubiquitous, the very idea of what a "ride" entails could transform. No longer just a mode of transportation, rides could be tailored experiences, offering anything from mobile workspaces to leisure pods equipped with entertainment and comfort amenities. Consumers could choose an experience rather than just a straightforward journey, and these customized services could be available at different tiers, further refining the variable ownership models.

In summary, the future sales and ownership models of vehicles will reflect more flexible, dynamic, and diverse approaches, driven by the capabilities of autonomous vehicles and shifting consumer preferences. With subscription services, car-sharing, corporate ownership, peer-to-peer leasing, modular updates, and more, the traditional notion of personal car ownership will give way to a new era of mobility. This transformative shift not only augments the convenience and accessibility of transportation but also sets the stage for a more sustainable and efficient future.

Chapter 18:
Legal Frameworks and Liability

The advent of autonomous vehicles introduces a complex web of legal frameworks and liability considerations that challenge traditional notions of responsibility on the road. While human drivers have historically borne the brunt of blame in accidents, the emergence of AI-driven cars necessitates a reevaluation of fault and accountability. Legislators and legal experts must grapple with questions such as: Who is liable when an autonomous vehicle is involved in a collision? Should it be the vehicle's manufacturer, the software developer, or perhaps the owner of the car? The evolving legal precedents are still nascent, with early cases serving as test beds for broader applications. As we navigate these uncharted waters, it's evident that creating fair and comprehensive legal standards will require international collaboration and a forward-thinking approach to ensure that both innovation and public safety are upheld.

Responsibility in Accidents

As we venture into the evolving landscape of autonomous vehicles, it's critical to understand where responsibility lies when accidents occur. The complexities multiply significantly when comparing human-driven cars to self-driving ones. Traditionally, humans have borne the burden of responsibility due to their direct control over the vehicle. But in a world driven by autonomous technology, responsibility becomes a far more intricate issue.

AI at the Wheel

The question of who is liable in the event of a collision involving an autonomous vehicle ignites considerable debate. Is it the manufacturer who failed to ensure the car's safety? Could it be the software developer who did not anticipate a particular scenario? Or is it the car owner, who might not have maintained the vehicle properly? Each case requires meticulous examination to sift through these possibilities, and current legal frameworks are often ill-equipped to deal with such nuances.

Novel legal theories are being tested in courtrooms around the world. For instance, product liability could take center stage. Under this premise, manufacturers would be held accountable if their vehicles are deemed defective, analogous to how liability works with other consumer products. However, this theory must be adaptable to encompass complex software failures and not just physical defects.

On the flip side, arguments are also being made for shared liability. Imagine a scenario where both a human driver and an autonomous vehicle contribute to an accident. This shared responsibility would require a fine-grained analysis of human actions vs. machine decisions. Even with advanced algorithms, machines may not always anticipate unpredictable human behavior, making this an exceptionally challenging area for legislatures and policymakers.

Additionally, the principle of contributory negligence comes into play. Suppose a person knowingly engages the autonomous function in an unsafe manner, despite warnings not to do so. In such cases, their contribution to the accident might reduce or limit their ability to claim damages. Determining contributory negligence would demand data from the vehicle's operating system, much like aviation black boxes, to reconstruct events preceding the accident.

Another angle to consider is software updating and maintenance liability. Autonomous vehicles are heavily dependent on regular software updates to improve safety and performance. If an accident happens because an owner failed to install these updates, questions of responsibility could shift significantly. The vehicle maker might argue that their software was safe, and the onus was on the owner to keep it updated, similar to how smartphone companies handle software vulnerabilities.

Insurance companies are also part of this conversation. They are developing new models to accommodate the shift from human to machine responsibility. Traditional car insurance schemes rely on assessing the risk posed by human drivers. As vehicles take on more driving tasks, insurers are redefining their risk assessment models. Premiums could be based more on the vehicle's software and hardware reliability rather than the driver's history, fundamentally changing the industry.

A significant complication arises when dealing with cross-jurisdictional issues. Different regions might have varied legal frameworks regarding autonomous vehicles, making it challenging to establish a universal standard for liability in accidents. International collaborations and harmonizing regulations are imperative to resolve these disparities, but achieving consensus is no small feat. Legal precedents set in one country may not be applicable in another, adding layers of complexity and uncertainty.

Lastly, the ethical dimension cannot be ignored. Developers of autonomous driving algorithms face moral dilemmas often encapsulated in the "trolley problem." Should the algorithm prioritize the lives of passengers over pedestrians or vice versa? These decisions have profound implications for determining responsibility. Legal systems must catch up with these ethical considerations, potentially leading to new forms of liability specific to the choices made by autonomous systems under life-and-death scenarios.

Overall, addressing responsibility in accidents involving autonomous vehicles demands an interdisciplinary approach, integrating insights from law, technology, ethics, and insurance. As autonomous vehicles become more prevalent, our society must evolve its legal frameworks and liability mechanisms to ensure they serve the public interest effectively. The journey will be complex, but establishing clear lines of responsibility is crucial for the future of autonomous transportation.

evolving legal precedents

As autonomous vehicles (AVs) become more integrated into our transportation networks, the legal landscape surrounding them continues to evolve. Legislators, judges, and regulators are constantly responding to new scenarios that arise from the deployment of AVs, crafting policies that balance innovation with public safety. This dynamic process is marked by a series of legal precedents that set the stage for future legislation and industry standards.

The historical foundation for many of today's legal precedents in autonomous driving can be traced back to early cases involving vehicular accidents and product liability. Traditionally, these cases hinged on the concept of human error. However, the introduction of autonomous technology complicates this aspect, raising questions about the extent of responsibility held by manufacturers, software developers, and even the vehicle itself. Courts are now faced with the challenge of determining liability in scenarios where machine learning algorithms play a pivotal role in decision-making.

One of the earliest notable cases involved a collision between an autonomous test vehicle and another car. The court's decision to hold the software company partially responsible marked a significant departure from the norms of human driver liability. This landmark ruling underscored the inherent complexities in distinguishing between human and machine fault. It also emphasized the need for clear guidelines that could help navigate these gray areas.

In another significant legal precedent, the question of ethical programming in AVs came into sharp focus. A case in which an autonomous vehicle had to decide between colliding with a pedestrian or swerve to hit another vehicle highlighted the moral implications of algorithmic decision-making. The court ruled that manufacturers must incorporate ethical decision-making frameworks into their software, adding an additional layer of responsibility for developers.

Legislation has also begun to adapt in response to these evolving precedents. Several states and countries have enacted laws that require AV companies to adhere to stringent safety standards, maintain detailed logs of software updates, and disclose the decision-making parameters embedded in their algorithms. These laws aim to ensure transparency and accountability, making it easier to assess liability when accidents occur.

Moreover, as part of a broader regulatory effort, some jurisdictions have established dedicated agencies or task forces focused on AV oversight. These bodies are tasked with analyzing data from real-world use cases, issuing guidelines, and providing a forum for stakeholders to discuss emerging legal issues. This approach aims to create a more responsive and adaptive legal framework that can keep pace with technological advancements.

Internationally, the picture is even more complex. While some countries have swiftly moved to embrace AV technology through favorable legislation, others have taken a more cautious stance, citing concerns about safety and data privacy. This divergence in legal approaches has led to a patchwork of regulations, making it challenging for manufacturers to navigate compliance across different regions. Nonetheless, international bodies like the United Nations Economic Commission for Europe (UNECE) are working towards harmonizing these regulations to foster a global standard.

An emerging trend is the collaboration between AV companies and insurance firms to develop new liability models. Traditional insurance policies are often ill-suited to cover the unique risks associated with autonomous driving. As a result, new types of coverage are being explored, such as policies that focus on software integrity, data security, and even "no-fault" frameworks that protect both manufacturers and consumers. These initiatives are likely to influence future legal precedents, further integrating the role of insurance in the AV ecosystem.

On a more granular level, local governments are also playing a crucial role in shaping legal precedents. Municipalities that serve as testbeds for AV technologies often have to draft tailor-made regulations to manage pilot programs. These localized efforts provide valuable case studies that can inform broader legal strategies. For example, cities that have allowed AVs to operate in mixed traffic conditions offer insights into pedestrian safety, traffic flow, and public acceptance, all of which are critical for crafting effective legal policies.

Public perception and societal values also influence legal precedents in the realm of autonomous driving. Courts and legislators are increasingly considering the ethical dimensions of AV deployment, balancing the benefits of innovation with potential societal risks. Public opinion can sometimes accelerate legislative action, especially in the wake of high-profile incidents involving AVs. The resulting legal decisions often set important precedents that guide future policy development.

Industry standards and best practices are another area where legal precedents are evolving. Regulatory bodies are working closely with industry leaders to develop comprehensive frameworks that can serve as benchmarks for AV safety and performance. These standards often become de facto legal requirements as courts and legislators look to them when making decisions. In this way, the interplay between regulation and industry practice helps to create a more cohesive legal landscape.

As legal precedents continue to develop, the role of interdisciplinary research cannot be overstated. Scholars from fields such as law, ethics, engineering, and computer science are collaborating to address the multifaceted challenges presented by autonomous driving. Their work often informs legislative and judicial decisions, contributing to a more robust and well-rounded legal framework. This collaborative approach ensures that legal precedents are grounded in a comprehensive understanding of both the technology and its societal implications.

Looking to the future, the legal landscape for autonomous vehicles will undoubtedly continue to evolve. Emerging technologies such as advanced machine learning, vehicle-to-everything (V2X) communication, and even quantum computing will introduce new legal challenges and opportunities. Policymakers will need to remain agile, adapting existing frameworks and creating new ones to address the complexities of these advancements. The dynamic nature of legal precedents in AVs reflects the broader trajectory of technological progress, marked by continuous innovation and adaptation.

Ultimately, the evolving legal precedents surrounding autonomous vehicles are a testament to the intricate relationship between technology and law. As AVs become an integral part of our transportation systems, the legal frameworks that govern them will continue to shape their development and deployment. By staying attuned to these legal developments, industry professionals, policymakers, and the public can better navigate the transformative impact of autonomous driving on society.

This intricate dance between technological innovation and legal regulation ensures that as we move forward, we do so in a manner that prioritizes safety, accountability, and ethical considerations. The journey of autonomous vehicles is as much a legal and societal one as it is a technological marvel, charting a course that requires careful navigation through the evolving landscape of legal precedents.

Chapter 19:
Insurance and Autonomous Vehicles

As autonomous vehicles continue to disrupt traditional driving paradigms, the insurance industry is undergoing its own transformation to adapt to these new technologies. Traditional car insurance models, which heavily rely on driver behavior and accident history, must evolve to incorporate advanced risk assessment techniques that account for AI-driven decision-making. Insurers are now developing algorithms to analyze data from vehicle sensors, software systems, and even environmental conditions to evaluate risk more accurately. This comprehensive data collection allows for new policy models that can offer dynamic premiums, adjusted in real time based on a vehicle's operating conditions and performance. Furthermore, the emergence of autonomous vehicles brings questions of liability to the forefront—who is at fault if an algorithm fails, the car manufacturer, software developer, or the vehicle owner? These evolving complexities require a nuanced approach both legally and financially, pushing the insurance sector to innovate rapidly to keep pace with the changes ushered in by autonomous driving technology.

New Models and Policies

As autonomous vehicles become a staple on our roads, it's crucial to revisit and, in many cases, completely overhaul existing insurance models. Traditional car insurance policies, designed around human drivers, no longer suffice in an era where artificial intelligence controls the wheel. This shift necessitates innovative insurance frameworks that address the nuances of autonomous driving while ensuring fairness and clarity for all stakeholders involved.

One of the first significant transformations in insurance models for autonomous vehicles is the reassessment of risk. In conventional vehicle insurance, risk is primarily determined by the driver's history, age, and behavior. However, with AI-driven vehicles, the focus shifts significantly. The emphasis now is on the software's reliability, the vehicle's technological sophistication, and the manufacturer's track record for safety. Insurers must grapple with questions of how to quantify and distribute risk when human intervention is minimal or entirely absent.

Another pivotal change involves liability. Traditionally, the driver is almost always at the center of liability discussions post-accident. However, with autonomous vehicles, determining fault becomes more complex. Is the accident a result of a software glitch, sensor failure, lack of proper maintenance, or a rare human intervention? To navigate this intricacy, new policies often incorporate clauses that hold manufacturers and software developers accountable under specific circumstances. This shift may also lead to hybrid models where part of the liability lies with the vehicle owner for maintenance and software updates, and part with the manufacturer for systemic failures.

Emerging alongside these changes is the concept of "usage-based insurance" (UBI). With UBI, insurers can offer more personalized premiums based on real-time data rather than demographic assumptions. For instance, a vehicle's sensors and connected systems can constantly relay information about driving patterns, environmental conditions, and vehicle usage to the insurer. Policies can then be dynamically adjusted to reflect the actual risk associated with the vehicle's operations, providing a fairer and more accurate premium calculation.

Regulatory bodies and governments play a crucial role in shaping and enforcing these new insurance models. As laws and regulations evolve to keep pace with technological advancements, a clear legal framework must be established to address disputes, accidents, and liabilities involving autonomous vehicles. Governments are also exploring the establishment of central databases to collect and analyze accident data from autonomous vehicles, facilitating more effective oversight and rapid policy adjustments.

Policy changes must also consider the broader societal implications of autonomous vehicle insurance. For instance, how do these new models affect individuals who may not be able to afford the latest technology? Will there be an increased risk of insurance redlining if certain autonomous systems are proven safer than others? To ensure equitable access to insurance and the benefits of autonomous driving, regulatory bodies might need to introduce subsidies or incentives for adopting safer, yet economically accessible technology.

Cybersecurity is another critical element intertwined with insurance policies for autonomous vehicles. With vehicles increasingly relying on sophisticated software and connectivity, the risk of cyber-attacks escalates. Insurers are requiring robust cybersecurity measures as part of their policies, ensuring that vehicles are protected against hacking attempts that could lead to accidents or system failures. This leads to new forms of coverage specific to cybersecurity threats, adding another layer of complexity to modern vehicle insurance.

Insurers are also forging partnerships with technology companies, sharing data and expertise to refine their models. These collaborations are central to understanding new risk factors and developing policies that fairly measure and manage these risks. For example, some insurance companies are teaming up with autonomous vehicle manufacturers to gain insights from collision data, helping them adjust premiums and coverages more accurately.

One significant implication of these new insurance models is the potential for reduced premiums over time. As autonomous vehicles are statistically less likely to be involved in accidents compared to human-driven cars, the overall claims should decrease. However, this is contingent on the technology's maturity and widespread adoption. In the early stages, premiums might be higher due to the high costs associated with repairing technologically advanced vehicles and the uncertainty surrounding new risks.

Furthermore, these insurance models must address the transition period where both autonomous and traditional vehicles share the road. Mixed traffic scenarios present unique challenges; insurers need to navigate the risk dynamics between autonomous and human-driven cars. Policies might need to include detailed protocols for accidents involving mixed interactions, ensuring that blame and subsequent claims are handled transparently and fairly.

Lastly, public awareness and education play a significant role in the successful implementation of new insurance models. Consumers must understand how these policies work, their benefits, and their responsibilities. Insurers should invest in educational campaigns, providing clear and accessible information to help consumers make informed decisions about their coverage options in the realm of autonomous vehicles.

The transition to autonomous vehicles necessitates a radical rethink of car insurance. New models and policies will need to adapt continuously, integrating technological advancements, regulatory changes, and societal needs. By staying ahead of these trends, insurance companies can not only remain relevant but also effectively contribute to the safe and equitable adoption of autonomous driving technologies.

Risk Assessment Techniques

In the evolving landscape of autonomous vehicles, insurance companies find themselves navigating uncharted waters. Traditional risk assessment models, rooted in human driver behaviors and historical accident data, must now adapt to the complexities of AI-driven systems. At its core, assessing risk for autonomous vehicles requires a holistic understanding of not just the technology itself, but also the dynamic interactions between machine, environment, and human factors.

First, let's consider the technological component. To accurately gauge the risk associated with autonomous vehicles, insurers must delve into the intricacies of sensors, algorithms, and hardware that govern these machines. Autonomous systems rely heavily on cameras, lidar, radar, and various other sensors to interpret their surroundings. Each of these components has its own failure modes and performance limits, which can vary based on environmental conditions such as weather and lighting. Therefore, insurance models must factor in the reliability and redundancy of these technologies to predict potential mishaps.

Moreover, the AI algorithms that drive these vehicles are often complex and opaque, making it difficult to pinpoint their decision-making processes. This introduces a layer of uncertainty that is not present in human-driven vehicles. Machine learning models, especially those based on deep learning, continuously evolve by learning from new data. While this adaptability improves functionality, it also means that the system's behavior can change over time in unexpected ways. To mitigate this risk, insurers need to collaborate closely with technology developers to gain insights into algorithmic changes and their potential impacts.

Environmental factors also play a critical role in risk assessment. Autonomous vehicles must navigate through ever-changing road conditions, traffic patterns, and pedestrian behaviors. Unlike human drivers who can rely on intuition, autonomous systems must process vast amounts of data in real-time to make split-second decisions. This dynamic environment introduces several risk variables that need to be assessed rigorously. For instance, factors like urban density, road infrastructure quality, and even local weather patterns can influence the likelihood of accidents. Insurers need to develop models that can incorporate these variables to provide a more accurate risk profile.

Human factors present another layer of complexity. While autonomous vehicles aim to reduce human error, they must still coexist with human drivers, pedestrians, and cyclists who may not always act predictably. It's essential to assess the interplay between human and machine, recognizing scenarios where human intervention might be necessary. Additionally, understanding public perception and the way people interact with autonomous vehicles can offer valuable insights into potential risks. For example, widespread acceptance and comfort with the technology could lead to smoother integration and lower risk, while skepticism or misuse could heighten it.

Another significant aspect is the analysis of historical data and real-world testing outcomes. Autonomous vehicles are subjected to extensive testing in simulations and controlled environments, but real-world performance can vary. Insurers must continuously analyze data from these vehicles to identify patterns and anomalies that might indicate underlying risks. By leveraging big data analytics and machine learning techniques, they can refine their risk models over time, improving accuracy and predictive capabilities.

On a broader scale, regulatory compliance and legal standards are fundamental in shaping risk assessment models. Autonomous vehicles operate within a complex web of regulations that differ across regions and countries. Insurers must stay abreast of these laws to ensure that their models are compliant and that they account for any legal liabilities that may arise. Regular updates to these models are necessary as regulations evolve and new legal precedents are established.

Innovation in risk assessment techniques is crucial for keeping pace with the rapid advancements in autonomous vehicle technology. One emerging approach is scenario-based simulation testing, which creates virtual environments that mimic a wide range of real-world scenarios. This allows for the assessment of how autonomous vehicles would respond to rare or extreme conditions that may not be frequently encountered in actual driving. Using these simulations, insurers can gain a better understanding of edge cases and develop more resilient risk models.

Furthermore, the integration of telematics data offers another valuable avenue for risk assessment. By capturing real-time data from vehicles, such as speed, braking patterns, and route choices, insurers can gain a detailed view of how these machines are being operated. This granular data enables more personalized risk profiles, which can lead to more accurate premium pricing and better risk management.

To bridge the gap between traditional risk assessment methods and the needs of autonomous vehicles, insurers should also consider adopting a collaborative approach. Partnerships with tech companies, academic institutions, and regulatory bodies can provide access to cutting-edge research and a deeper understanding of emerging risks. Such collaboration can lead to the creation of standardized frameworks and best practices, fostering a more cohesive and comprehensive approach to risk assessment.

Lastly, transparency and communication with consumers will play a pivotal role. As autonomous vehicles become more prevalent, educating consumers about the risks and coverage options will be essential. Clear communication can help build trust and ensure that consumers are making informed decisions about their insurance needs.

In conclusion, the advent of autonomous vehicles necessitates a fundamental shift in how risk is assessed within the insurance industry. By embracing innovative technologies, continuously analyzing data, and fostering collaboration, insurers can develop robust models that accurately reflect the complexities of this new era of transportation. The road ahead may be filled with challenges, but with the right strategies, insurers can navigate these new frontiers with confidence.

Chapter 20:
AI in Navigation and Mapping

As autonomous vehicles inch closer to becoming a common sight on our streets, the role of AI in navigation and mapping has become pivotal. Real-time updates enhance the accuracy of maps, allowing self-driving cars to make split-second decisions based on current road conditions, traffic, and hazards. Advanced routing algorithms continually optimize paths, balancing efficiency with safety. These algorithms rely on vast datasets and machine learning to predict traffic patterns and even adapt to unforeseen changes in the environment. Ultimately, AI-driven navigation systems are not only making self-driving cars more reliable but also setting new standards for human-centric transportation, pushing the boundaries of what modern maps and routing can accomplish.

Real-Time Updates and Accuracy

In an era where traffic congestion and navigation inefficiency are constant hurdles, AI's transformative role in providing real-time updates and ensuring mapping accuracy is revolutionary. One of the critical elements powering autonomous vehicles is the capability to adapt instantaneously to dynamic road conditions. This capacity hinges heavily on the precision and prompt delivery of real-time data.

The cornerstone of real-time updates in navigation is a robust network of sensors and data collection points. Autonomous vehicles are equipped with a myriad of sensors, including LiDAR, radar, and cameras, that constantly scan the environment. This sensory data is augmented with information from external sources such as satellites, traffic cameras, and even other connected vehicles. The fusion of these numerous data streams enables an autonomous vehicle to maintain a comprehensive, real-time understanding of its surroundings.

But generating data alone isn't sufficient. The data must be processed and interpreted rapidly to be actionable. Advanced machine learning algorithms come into play here, employing techniques like deep learning and neural networks to sift through terabytes of information in milliseconds. These algorithms analyze and predict potential traffic scenarios, road hazards, and even pedestrian movements, thereby offering immediate, accurate updates to the vehicle's navigation system.

A pivotal challenge in maintaining real-time accuracy is the variability and unpredictability of road conditions. Accidents, construction zones, abrupt weather changes, and other unforeseen events can significantly impact navigation. This is where AI excels over traditional GPS systems. With continuous learning and adaptation, AI-driven systems can re-route vehicles in real-time, guiding them away from traffic bottlenecks and towards the most efficient paths.

Moreover, AI in navigation isn't just about reactive measures; it's also about predictive capabilities. Through historical data analysis and pattern recognition, AI systems can forecast traffic trends and potential obstacles before they occur. For instance, by analyzing rush hour patterns, these systems can recommend optimal departure times or suggest alternative routes that minimize delays.

Human error remains a significant concern in traditional navigation systems. Mistakes in manual map updates, delayed information relay, and the reliance on outdated maps can lead to inefficiencies. AI mitigates these issues by providing a continuous, automated update mechanism. These systems leverage continuous crowdsourced data, ensuring that maps are always current. Whether it's a newly constructed road, a temporary detour, or changes in traffic regulations, AI systems integrate these elements promptly, enhancing navigation accuracy.

Real-time updates also contribute significantly to the safety framework of autonomous vehicles. Rapid-response algorithms can detect and react to potential hazards almost instantaneously. For instance, if there's a sudden roadblock or an impending collision threat, the AI system can execute an evasive maneuver in fractions of a second, safeguarding passengers and pedestrians alike.

The integration of AI in navigation systems extends beyond individual vehicles. The concept of Vehicle-to-Everything (V2X) communication plays a crucial role in creating a synchronized traffic ecosystem. V2X allows vehicles to exchange information with other vehicles (V2V), traffic signals (V2I), and even pedestrians via smart devices (V2P). This interconnected system amplifies the accuracy and timeliness of real-time updates, creating a cohesive network where each participant contributes to and benefits from the shared data.

Another fascinating application is the role of AI in enhancing user experience during navigation. Traditional GPS systems often provide a static list of instructions, but AI in navigation offers a more interactive experience. Voice-enabled AI assistants can give context-aware suggestions and answer on-the-fly queries about road conditions, detours, and estimated arrival times. This not only improves accuracy but also enhances driver trust in the system.

Let's not overlook the implications for ride-sharing and public transportation. For instance, ride-sharing platforms benefit immensely from real-time updates to optimize fleet deployment. By directing drivers to the busiest areas at specific times, these services can reduce wait times and improve user satisfaction. Likewise, public transit systems can employ AI to monitor and manage real-time schedules, leading to more reliable services. In both scenarios, the accuracy of AI-driven navigation enhances operational efficiency and user experience.

The continual evolution of 5G technology further bolsters the efficacy of real-time updates. With its low latency and high-speed data transfer capabilities, 5G ensures that the vast amounts of data required for real-time navigation are transmitted and processed without delay. As 5G networks become more widespread, the reliability and scope of AI-driven real-time updates will only improve, paving the way for even more sophisticated autonomous driving systems.

Real-time updates also play an essential role in emergency response scenarios. Autonomous vehicles deployed in emergency services can receive real-time data to avoid congested routes, arrive at the scene faster, and potentially save lives. Hospitals, too, can benefit from this technology by coordinating arrivals and departures in busy urban environments, ensuring that emergency vehicles have a clear path.

In essence, the integration of real-time updates and accuracy through AI drives the next wave of innovation in navigation and mapping. By constantly adapting to an ever-changing environment, these systems not only make autonomous vehicles more efficient but also safer and more reliable. As we continue to leverage advancements in AI, machine learning, and data collection, the full potential of real-time updates will undoubtedly transform how we approach navigation and transportation, offering unprecedented levels of accuracy and reliability.

Advanced Routing Algorithms

Advanced routing algorithms form the backbone of AI in navigation and mapping. These algorithms are at the heart of how autonomous vehicles determine the most efficient, safe, and reliable paths from point A to point B. Routing isn't just about finding the shortest path; it's about considering factors like traffic conditions, road types, weather challenges, and even pedestrian activities. All these elements demand sophisticated, dynamic algorithms that can adapt in real-time.

One of the most common algorithms in traditional routing is Dijkstra's Algorithm, known for finding the shortest path in a graph. Though revolutionary in its time, it falls short when applied to the complex and dynamic nature of real-world road networks. Modern autonomous vehicles require algorithms far more advanced and nuanced. These advanced algorithms are often hybrids, incorporating elements from Dijkstra's but enhancing them with machine learning techniques to create more adaptable and smarter routing strategies.

These advanced algorithms take advantage of big data analytics. Massive datasets collected from GPS systems, traffic cameras, and even social media feeds provide real-time insights into road conditions. Machine learning models can analyze this flood of data to predict traffic trends and identify potential hazards. The blend of historical data and real-time updates allows these algorithms to make highly accurate routing decisions that are far superior to traditional methods.

Consider the role of predictive analytics. By using AI to analyze patterns in traffic flow, weather conditions, and even seasons, predictive algorithms can anticipate future conditions rather than just react to existing ones. This foresight is invaluable in making proactive routing decisions. For example, if an algorithm predicts a traffic jam due to an upcoming football game, it can reroute vehicles well in advance, saving time and improving the driving experience.

One prominent technique employed in these advanced routing algorithms is reinforcement learning. This method involves training an AI model to make a series of decisions that maximize a particular reward, such as minimizing travel time or fuel consumption. Over time, the model learns to navigate various scenarios more efficiently by 'learning' from its mistakes and successes. Reinforcement learning is particularly useful for autonomous vehicles because it can adapt to new and unforeseen conditions, improving its routing strategy continuously.

Graph Neural Networks (GNNs) represent another cutting-edge approach in the development of routing algorithms. GNNs treat road networks as graphs, where intersections are nodes and roads are edges. These networks can process complex relational data, allowing algorithms to understand intricate patterns and dependencies in the road network. The result is a routing system that doesn't just look at individual segments of a journey but considers the entire network holistically.

AI-driven routing algorithms also integrate vehicle-to-everything (V2X) communication technologies. V2X allows vehicles to communicate with each other and with infrastructure elements like traffic lights and road signs. This communication enables real-time data exchange, which is fed back into the routing algorithms. Imagine a scenario where a vehicle receives a signal from a traffic light about to turn red; the algorithm can use this data to optimize the route, avoiding delays and improving efficiency.

Aside from efficiency and time-saving, these advanced algorithms prioritize safety. Route planning involves assessing the quality of roads, the likelihood of encountering pedestrians, and the probability of running into hazardous conditions. AI models can evaluate these factors and choose routes that minimize risks. This level of advanced planning significantly enhances the reliability of autonomous driving systems.

However, the development of these algorithms is not without challenges. Real-world conditions are immensely variable. An algorithm might perform superbly in one city but falter in another due to differences in road infrastructure, driving behavior, and even laws.

Regional customization of algorithms is essential but resource-intensive. Researchers and engineers must continuously test and refine these models in various settings to ensure their robustness.

Open-source platforms also play a crucial role in the development and dissemination of advanced routing algorithms. Initiatives like OpenStreetMap provide valuable data and tools that researchers and developers can use to train their models. Collaborative platforms enable rapid innovation by democratizing access to data and fostering a community of shared knowledge.

Regulatory standards are another critical aspect. For advanced routing algorithms to be effective and widely adopted, they must conform to legal requirements and standards set forth by transportation authorities. Regulations surrounding data privacy, real-time communication, and algorithmic transparency must be adhered to, ensuring that these systems are both effective and compliant.

As quantum computing advances, its potential to revolutionize routing algorithms can't be overlooked. Classical computing methods may struggle with the complexities and sheer volume of data involved in real-time, dynamic routing. Quantum algorithms, however, could solve these problems exponentially faster. While quantum computing is still in its infancy, its future applications in routing algorithms are a tantalizing prospect, promising improvements in speed and efficiency that are currently unimaginable.

The integration of IoT (Internet of Things) devices into routing algorithms further fine-tunes and optimizes routing decisions. Sensors embedded in roads, smart traffic lights, and even wearable devices can provide continuous data streams. This information enriches the algorithms, making them even more responsive and adaptable. For instance, sensors detecting icy road conditions can instantly relay this data, allowing the algorithm to reroute the vehicle, ensuring safety.

Lastly, public engagement and acceptance play a consequential role in the effectiveness of these systems. Users must trust that the routes suggested by these algorithms are not only the most efficient but also the safest. Transparency in how these algorithms operate, coupled with robust validation and testing protocols, can build this trust. Ultimately, the success of advanced routing algorithms hinges on their ability to deliver on their promises while gaining public confidence.

Advanced routing algorithms signify a remarkable leap towards highly functional autonomous navigation systems. By combining traditional principles with cutting-edge AI, these algorithms offer unprecedented levels of efficiency, safety, and reliability. As technology continues to evolve, so too will these algorithms, continually pushing the boundaries of what's possible in autonomous driving and navigation.

Chapter 21:
Human-Machine Interaction

In the dynamic realm of autonomous driving, the interaction between humans and machines is pivotal. The design of user interfaces must be intuitive enough for anyone to navigate, while also sophisticated enough to handle complex driving tasks. As AI continues to evolve, creating a sense of trust and comfort in users becomes essential. Addressing this involves not just technological advancements but also understanding human psychology and behavior. Personalized settings, real-time feedback, and adaptive learning systems can ease the transition, making riders feel more secure. Ultimately, enhancing human-machine interaction is key to the widespread acceptance and success of autonomous vehicles, bridging the gap between cutting-edge technology and everyday usability.

User Interface Design

User interface design in the context of autonomous driving is a crucial aspect of Human-Machine Interaction. As these vehicles become more prevalent, the way users interact with them must be intuitive, efficient, and safe. The design of user interfaces (UI) in self-driving cars revolves around ensuring that human users can easily understand and control the vehicle when necessary.

At the core of UI design in autonomous vehicles is the concept of user-centered design. Designers need to consider the various scenarios in which a user might need to interact with the car. For instance, while the vehicle is driving autonomously, the user might want to check the route, adjust the destination, or take over control. This necessitates interfaces that are simple, straightforward, and capable of quick response.

One of the primary challenges in UI design for autonomous vehicles is reducing cognitive load. Users should not be overwhelmed with too much information at once. Instead, the information should be presented in a way that is easy to process. Heads-up displays (HUDs) and augmented reality (AR) dashboards are emerging as effective tools in this regard. These technologies can overlay crucial information directly onto the windshield, allowing users to keep their eyes on the road while still receiving necessary updates.

The interaction between users and autonomous systems must also prioritize safety. For instance, if an emergency situation arises, there should be clear, unambiguous signals prompting the user to take control. This could involve auditory alerts, haptic feedback, and visual cues that are immediately recognizable and understood.

Moreover, as vehicles become more autonomous, the role of the human driver shifts from being an active controller to a passive monitor. This change requires an evolution in how interfaces communicate status and updates to the user. Regular status updates and the ability to easily switch between autonomous and manual driving modes are critical elements.

Another important aspect is inclusivity and accessibility in UI design. Autonomous vehicles should cater to a broad spectrum of users, including those with disabilities. Voice control systems, customizable interfaces, and adaptable displays can provide a more inclusive experience. Ensuring that the interface is usable by people with various physical and cognitive abilities is not just a legal requirement but a moral imperative.

The aesthetic element of UI design should not be underestimated either. A well-designed interface that is visually pleasing can significantly enhance the user experience. Consistent with the vehicle's overall design ethos, the interface should blend seamlessly with the car's interior. High-resolution screens, intuitive touch controls, and elegant graphics can make the UI not only functional but also enjoyable to use.

Interactivity is another cornerstone of effective UI design. For a seamless user experience, the interface must be responsive and fast. Long lag times or unresponsive controls can lead to frustration and mistrust. Therefore, optimizing the software behind the interface for speed and reliability is as important as the visual design.

Trust is another critical factor that UI designers must address. Users need to trust that the autonomous system will perform reliably. Clear, transparent communication about what the vehicle is doing and why can build this trust. Features like real-time feedback, journey progress updates, and transparent decision-making processes are essential.

Personalization plays a significant role as well. Allowing users to customize their interface can make the experience more comfortable and personal. From preferred routes to climate control settings, personalized interfaces can remember user preferences and adjust settings accordingly, providing a more tailored experience.

User testing and iterative design are indispensable in developing a successful UI. Gathering feedback from real users through beta testing phases can uncover unforeseen issues and areas for improvement. Continuous refinement based on user input ensures that the interface evolves to meet the needs and expectations of its users.

The evolution of UI in autonomous vehicles doesn't occur in isolation. It requires collaborative efforts from designers, engineers, psychologists, and other specialists. Integrating multidisciplinary insights ensures that the final product is well-rounded and effective, serving its purpose in a real-world context.

User interface design is pivotal in shaping how people perceive and interact with autonomous vehicles. A well-thought-out interface can make the difference between a smooth, enjoyable ride and a frustrating experience. Designers must juggle aesthetics, functionality, and safety while also retaining a focus on user needs and preferences. Only through such a balanced approach can the true potential of autonomous driving be realized.

Improving Trust and Comfort

In the advancement of autonomous vehicles, fostering trust and ensuring comfort for users remain central challenges. As we experience the steady integration of autonomous driving into our lives, prioritizing human elements such as trust and comfort can determine how smoothly this transition occurs. Interaction between humans and machines must be seamless and intuitive to earn the public's confidence and make the experience enjoyable.

Trust is often rooted in predictability and familiarity. When people step into a traditional car, they have an inherent understanding of its operation and expectations for its performance. Autonomous vehicles, on the other hand, represent a significant departure from this norm. To bridge this gap, developers need to focus on building interfaces and systems that users find logical and reassuring. This could include clear and straightforward communication from the vehicle about its actions and decisions, akin to a transparent and predictable human driver.

Communication is crucial in building this trust. Visual, auditory, and even tactile feedback can help inform passengers about the car's intentions. For instance, visual displays indicating upcoming actions—such as a lane change or an intended stop—can help passengers feel more in control and less anxious. Simple auditory cues can further reinforce this, providing an additional layer of comfort by mimicking the assurances one might receive from a human driver.

Comfort, both physical and psychological, is another key component. Physically, the design of autonomous vehicles should prioritize ergonomic seating, climate control, and smooth ride quality. Psychological comfort, however, is more nuanced and involves ensuring that passengers feel safe and secure within the vehicle. For instance, managing acceleration, deceleration, and turns in a way that minimizes abrupt movements can significantly enhance the feeling of safety.

One innovative approach involves adaptive systems that learn and adjust to the preferences of individual users. By leveraging AI, these systems can tailor the driving style to match the comfort levels of different passengers. For example, some users might prefer a more dynamic driving experience, while others may favor a gentler approach. By learning and adapting over time, these systems can provide a customized experience that enhances both trust and comfort.

Moreover, advanced machine learning algorithms can predict and respond to situations in a manner that mirrors human intuition. By anticipating potential conflicts or dangers on the road and taking preemptive action, autonomous vehicles can demonstrate a level of foresight that reassures passengers of their safety. This proactive approach to driving helps build trust, as users can see the vehicle handling complex scenarios with apparent ease.

Another aspect of improving trust and comfort lies in addressing edge cases and unexpected events. Autonomous vehicles must be equipped to handle a wide range of scenarios, from inclement weather to sudden obstacles. Rigorous testing and continuous refinement of

AI models are essential in preparing these vehicles for real-world conditions. Providing transparency into how these systems are tested and validated can also help build public confidence.

Social acceptance of autonomous vehicles is further enhanced when there is public education and awareness. By demystifying the technology and demarcating clear lines of capabilities and limitations, stakeholders can foster a more informed and trusting public. Workshops, demonstrations, and pilot programs can serve as valuable touchpoints for potential users to experience and understand the technology first-hand.

Building trust and comfort isn't a one-time effort but an ongoing commitment. Continuous feedback loops from users should inform iterative improvements. By constantly refining the systems based on real-world data and user experiences, developers can ensure that the technology evolves to better meet the needs of its users.

Another cornerstone is the vehicle's capacity for emergency response. In critical situations, the vehicle must be able to communicate effectively and may need to provide manual overrides. Ensuring that passengers know how to interact with these emergency systems can alleviate concerns and give them a sense of reassurance.

Collaboration between technology developers, regulatory bodies, and consumer advocacy groups can play a significant role in enhancing trust and comfort. By establishing and adhering to rigorous safety standards and transparency in reporting, stakeholders can help cultivate a climate of trust. For instance, standardized safety ratings for autonomous vehicles could become a key metric for consumers when evaluating their options.

In summary, improving trust and comfort in the context of human-machine interaction in autonomous driving is multifaceted. Effective communication, both in terms of the vehicle's intent and educational outreach to the public, forms the bedrock. Physical and psychological comfort need to be continuously addressed through innovative designs and adaptive systems. Rigorous testing, transparency, and collaborative efforts round out the comprehensive strategy required to usher in an era where autonomous vehicles are not only reliable but also embraced and trusted by society at large.

Chapter 22:
Autonomous Commercial Vehicles

As the capabilities of autonomous driving technology progress, the commercial sector is witnessing a transformative shift. Autonomous commercial vehicles, particularly trucks and delivery systems, are poised to revolutionize logistics and supply chains. These vehicles promise not only increased efficiency but also potential cost reductions and enhanced safety on highways. For businesses, this technology means fewer disruptions and more consistent delivery times. However, the transition isn't without its challenges; the economic impact on the trucking industry workforce and the necessary infrastructure upgrades are significant barriers. As we venture further into this autonomous era, balancing these benefits and challenges will be crucial for sustainable growth.

Trucks and Delivery Systems

Autonomous commercial vehicles are not just about passenger transport; they're about transforming the backbone of our economy: logistics and delivery systems. The idea of trucks and delivery vans driving themselves may sound futuristic, but it's becoming a reality faster than most anticipate. This section delves into how autonomous technology is reshaping the landscape of heavy-duty transportation and last-mile deliveries.

The first major shift in the autonomous trucks arena is the move towards long-haul trucking. Unlike city driving, interstate routes are more predictable, offering a less complex environment for implementing self-driving technology. Imagine a convoy of trucks moving in sync, controlled by sophisticated algorithms that can adjust their speed and positioning to optimize fuel consumption and reduce traffic congestion. This form of 'platooning' not only increases efficiency but also lowers operational costs.

With the growing e-commerce sector, the demand for efficient last-mile delivery has skyrocketed. Autonomous delivery vans are designed to navigate urban landscapes, delivering packages right to our doorsteps. These smaller vehicles come equipped with advanced AI systems capable of recognizing obstacles, detecting pedestrians, and interpreting traffic signals. They promise faster deliveries and can work round-the-clock, unlike human drivers.

So, how does this affect the operational landscape? Companies can now optimize their logistics networks to a level previously unimaginable. By integrating autonomous trucks with sophisticated fleet management software, they can predict delivery times with high accuracy, reduce fuel expenses, and decrease vehicle downtime. Such precision in operations translates to greater customer satisfaction and a significant competitive edge.

Nevertheless, rolling out a fleet of autonomous trucks and delivery vans involves overcoming numerous challenges. One of the primary concerns is regulatory approval. Each region has its own set of rules governing the operation of autonomous vehicles, and logistics companies must navigate this intricate legal landscape before deploying their fleets. Then there's the matter of safety. Companies must ensure these vehicles adhere to the highest safety standards to gain public trust and regulatory approval.

Another crucial aspect is the integration of autonomous systems with existing logistics infrastructure. While startups and tech giants are making strides in self-driving technology, they must work closely with traditional logistics operators to integrate these innovations smoothly. This cooperation ensures a seamless transition from human-operated to autonomous delivery systems.

The economic implications of autonomous trucks and delivery systems are profound. On the one hand, companies stand to save billions in labor, fuel, and maintenance costs. On the other hand, the shift could disrupt the livelihood of millions of truck drivers. However, new job opportunities will likely emerge, requiring skills in robot maintenance, AI system monitoring, and fleet management. This transformation demands a workforce that is adaptable and willing to learn new skills.

There's also an environmental angle to consider. Autonomous trucks are often more fuel-efficient and can adopt electric powertrains, significantly reducing emissions. Platooning, for instance, can cut down on aerodynamic drag, thus saving fuel. Likewise, electric delivery vans produce zero tailpipe emissions, making urban centers cleaner and greener.

Competition is heating up in the autonomous delivery space, with traditional vehicle manufacturers, tech firms, and startups racing to gain an edge. Companies like Tesla, Waymo, and Embark are pushing the envelope with innovative solutions tailored to different aspects of commercial transportation. Investment and mergers in this sector are also on the rise, indicating strong belief in the potential of autonomous logistics and delivery systems.

How soon will we see widespread adoption of these technologies? While fully autonomous commercial fleets may still be years away, incremental advancements are already evident. Semi-autonomous systems, where a human driver oversees operations but the vehicle handles most of the driving, are already in use. These systems serve as a transitional phase, paving the way for fully autonomous logistics.

In summary, autonomous trucks and delivery systems promise to revolutionize how goods are transported. The technology is rapidly evolving, offering solutions that are more efficient, cost-effective, and environmentally friendly. While challenges remain, the industry is poised for significant transformation. As autonomous vehicles become more prevalent, they'll reshape the logistics landscape, driving innovation and setting new standards for efficiency and sustainability.

Economic Benefits and Challenges

When delving into the economic benefits and challenges of autonomous commercial vehicles, it's essential to recognize the powerful impact these innovations are poised to have on a global scale. First and foremost, these vehicles promise to revolutionize the logistics and transportation sector. The potential for cost savings is significant, as autonomous trucks and delivery systems can operate around the clock without the need for rest, dramatically increasing efficiency and reducing downtime. Reducing the need for human drivers could also lead to substantial labor cost savings for companies, which in turn can make products and services cheaper for consumers.

However, these economic benefits come with significant challenges. One of the primary concerns is the displacement of jobs. The trucking industry employs millions of drivers worldwide. The shift to autonomous vehicles means many of these jobs may become obsolete, leading to widespread unemployment unless new roles are created within the industry, such as maintenance, monitoring, and technology management positions. Policymakers and industry leaders must address this potential displacement through thoughtful planning and workforce re-training initiatives.

The reduction in operational costs doesn't just come from labor savings. Autonomous vehicles can optimize fuel consumption through more efficient driving patterns and reduce wear and tear on the vehicles by avoiding sudden stops and starts. These efficiencies can translate into lower fuel and maintenance costs. There's also the prospect of reducing the number of accidents caused by driver error, potentially lowering insurance premiums and vehicle repair expenses. Thus, the integration of autonomous commercial vehicles could streamline operations and enhance profitability for businesses.

On the flip side, the economic landscape faces a significant challenge in terms of the initial investment required to transition to autonomous fleets. The technology itself, while advancing rapidly, remains expensive. Companies must consider the costs associated with upgrading their current infrastructure to accommodate these new systems, installing necessary hardware, and ensuring cybersecurity measures are up to par. For smaller companies, this financial hurdle could be particularly daunting, potentially widening the gap between large enterprises and SMEs in the industry.

Furthermore, the development and deployment of autonomous commercial vehicles necessitate a robust legal and regulatory framework. Establishing these frameworks involves significant financial and bureaucratic efforts, which can be both time-consuming and costly. Building trust and ensuring public safety means conducting rigorous testing phases and ongoing validation. The resources allocated towards compliance with these regulations could strain budgets, especially for emerging companies trying to make a mark in the market.

A secondary but important benefit lies in potential environmental savings. Autonomous vehicles have the capacity to reduce carbon footprints by optimizing routes and improving fuel efficiency. This shift could help companies align with stricter environmental regulations and meet sustainability goals, which are becoming increasingly pivotal in the global market. Companies that can minimize their environmental impact may benefit from government incentives and a better public image, creating additional economic advantages.

Nevertheless, the dependence on high levels of connectivity and advanced infrastructure presents another economic challenge. Rural areas, which may lag in terms of technological infrastructure, could find it difficult to capitalize on the benefits of autonomous commercial vehicles. Bridging this urban-rural divide will require substantial investment in upgrading infrastructure, which poses a significant economic challenge for both public and private sectors.

There's also the matter of cybersecurity risks that come with the introduction of autonomous technology. Ensuring security for vehicular networks requires continuous, costly updates to protect against cyber threats. Companies will need to invest heavily in cybersecurity measures to safeguard their fleets from potential attacks, which could lead to operational disruptions and financial losses. This exponentially increases the operational expenses and can diminish the economic benefits originally forecasted.

Lastly, public perception and market adaptation are crucial for the economic success of autonomous vehicles. Consumer trust in the technology and willingness to accept autonomous deliveries plays a key role. Overcoming public skepticism and ensuring widespread adoption requires educational campaigns, demonstrations of safety, and transparent communication about the technology and its benefits. This demand for social acceptance carries its own economic implications, as companies might need to invest in comprehensive marketing and public relations strategies.

In summary, the economic benefits of autonomous commercial vehicles are compelling, encompassing cost savings, increased efficiency, and environmental gains. However, these benefits are tempered by significant challenges such as job displacement, high initial investment costs, regulatory hurdles, infrastructure demands, cybersecurity threats, and the need for public acceptance. Addressing these challenges will require collaborative efforts between industry leaders, policymakers, and communities to navigate the transition towards an autonomous future successfully. While the road ahead is complex, the potential for transformative economic impacts makes it a journey worth undertaking.

Chapter 23:
The Role of Startups and Innovators

Startups and innovators are the lifeblood of the autonomous vehicle revolution, driving forward disruptive technologies that push the boundaries of what's possible. These small, agile companies often introduce groundbreaking ideas that might be overlooked by more established automotive giants. With an increasing number of venture capitalists pouring millions into the autonomous driving sector, these startups have the financial backing to turn innovative concepts into reality. They focus on niche problems, such as enhanced sensor technologies or more efficient machine learning algorithms, contributing to the intricate puzzle of fully autonomous driving. Not bound by the same bureaucratic constraints as larger corporations, startups can experiment, pivot, and iterate quickly, providing essential breakthroughs that set new industry standards. Their role isn't merely supplemental; it's foundational, ensuring continuous advancement and robust competition in a rapidly evolving field.

Disruptive Technologies

The realm of autonomous driving is undergoing a seismic shift, thanks to a slew of disruptive technologies that startups and innovators are bringing to the forefront. At its core, disruption means upending traditional ways of thinking and doing; it means new paradigms of efficiency and connectivity previously unimagined. It's not just the autonomous vehicles themselves that matter, but the technologies and ideas that power them.

Consider Lidar and radar systems. These technologies have seen significant enhancements, driven by startups that have developed more cost-effective, smaller, and more accurate sensors. By innovating in Lidar technology, companies like Velodyne and Luminar have made it possible for autonomous vehicles to have a "sixth sense," mapping their environment in real-time with previously unattainable precision. This, in turn, allows for better decision-making processes within the vehicle's AI systems.

Artificial Intelligence and machine learning are, arguably, at the heart of these disruptive technologies. Startups specializing in AI-driven algorithms have taken giant strides in predictive analytics and real-time data processing. These advancements allow vehicles to "learn" from each trip, optimizing routes, and improving safety measures. AI isn't just powering navigation; it's making the entire system more intuitive, efficient, and responsive.

Connectivity is another major area where startups are making waves. Vehicle-to-Everything (V2X) communication is becoming more crucial as autonomous vehicles need to interact not just with each other but with infrastructure like traffic lights and road signs. Innovators are developing highly reliable communication protocols that make this possible. Technologies such as 5G and edge computing are fundamental here. They enable near-instantaneous data transfer, facilitating real-time decision-making and ensuring a seamless flow of information between all parts of the system.

Battery technology is also witnessing remarkable innovation, mainly led by startups focused on alternative energy sources. These companies are coming up with lighter, longer-lasting batteries, making electric autonomous vehicles more viable and efficient. The move towards solid-state batteries, for instance, promises greater safety and faster charging times. Breakthroughs in this sector imply that the dream of completely electric, high-performance autonomous vehicles is closer than ever.

In addition to hardware and connectivity, disruptive software solutions are transforming how autonomous vehicles operate. Companies are developing advanced simulation software that allows for millions of miles' worth of road tests to be conducted virtually. This approach drastically speeds up the validation process for self-driving systems, ensuring they are safe and reliable before they even hit the roads.

The Internet of Things (IoT) can't be overlooked when discussing disruptive technologies. IoT enables vehicles to become part of a broader, interconnected ecosystem. Startups are capitalizing on this by developing IoT platforms that enable vehicles to communicate with smart city infrastructures, homes, and even other personal devices. This connectivity extends the utility of autonomous vehicles far beyond transportation, integrating them into our daily lives in ways we are just beginning to understand.

Another area rife with disruption is human-machine interface (HMI) technology. As autonomous vehicles become more prevalent, the way humans interact with them will become increasingly important. Startups are experimenting with intuitive HMI solutions that use voice commands, haptic feedback, and even augmented reality to create a more user-friendly experience. These interfaces aim to make the transition from manual to autonomous driving smooth, enhancing user trust and acceptance.

Cybersecurity is another critical area witnessing significant disruption. With the increase in connectivity and data exchange, the threat landscape has become more complex. Innovators are developing robust cybersecurity measures to protect against hacking and data breaches. Companies are focusing on creating secure communication protocols and fail-safe systems that ensure the vehicle can operate safely even in the face of attempted cyber-attacks.

The investment landscape has also evolved dramatically. Venture capital is pouring into these disruptive technologies, with investors keen to get in on the ground floor of what promises to be a revolutionary industry. Government grants and partnerships are further catalyzing progress, particularly for startups who might lack the capital to push their innovations to market maturity. Crowdfunding platforms are another avenue through which disruptive ideas are finding the financial support they need.

Open-source platforms are playing a crucial role in driving innovation. These platforms allow startups to build upon existing technologies without incurring prohibitive costs. By contributing to and borrowing from a shared pool of knowledge, smaller companies can innovate more rapidly and efficiently. These collaborative efforts mean that technological advancements are not siloed but benefit the entire industry.

Furthermore, urban mobility startups are radically reimagining transportation networks with autonomous vehicles at the core. They are experimenting with various business models, from fleet-based ride-sharing services to autonomous delivery systems. These models could drastically reduce traffic congestion and offer more sustainable urban transport options, aligning with broader societal goals of reducing carbon emissions and enhancing public transportation.

Innovation is not limited to software and hardware but extends to governance and policy-making as well. Regulatory technology, or "Regtech," is emerging as a catalyst for change, helping startups navigate complex legal landscapes more efficiently. These technologies aid in real-time compliance monitoring and reporting, enabling faster adaptation to evolving laws and guidelines.

All these disruptive technologies collectively contribute to a future where autonomous driving isn't just a concept but a well-integrated reality. They represent a paradigm shift that promises to redefine our relationship with transportation, making it more efficient, safer, and deeply interconnected with our day-to-day lives. The ongoing efforts of startups and innovators are not just incremental improvements but quantum leaps that have the potential to upend existing transportation paradigms.

Investment and Funding Landscape

As the potential for autonomous driving continues to capture the imagination of both the public and investors, the investment and funding landscape surrounding startups and innovators in this space has become increasingly dynamic. The financial ecosystem supporting these ventures is intricate, involving a myriad of stakeholders, each playing a pivotal role in pushing the boundaries of what's possible.

Venture capital, often considered the lifeblood of high-tech innovation, has become particularly enamored with the possibilities inherent in autonomous vehicles (AVs). Early-stage ventures, bursting with disruptive ideas yet requiring substantial financial backing to achieve their goals, frequently turn to venture capital firms for support. These firms, in turn, are drawn by the potential for massive returns, driven by the expectations of AVs revolutionizing transportation, logistics, and urban planning.

Angel investors also constitute a significant portion of initial funding for autonomous vehicle startups. These individual investors are typically keen on investing early, even before the product or service has been fully validated. The risks are high, but so are the potential rewards. Successful exits via acquisitions or IPOs have made angel investing a lucrative entry point into the burgeoning AV sector.

AI at the Wheel

The dual forces of excitement and caution have shaped the landscape of corporate venture capital (CVC) as well. Major automotive companies such as Ford, GM, and Toyota have established their own investment arms, pumping money into promising startups that can bring fresh technology and competitive advantages. This strategic investment not only secures future innovation but also provides these corporations a seat at the table in discussions about future mobility.

Strategic partnerships are also a common vehicle for funding, merging the financial and technical resources of large firms with the agility and novel ideas of startups. Collaborations between tech giants like Google or Nvidia and smaller AV developers enable the integration of cutting-edge technologies and provide the startups with access to a broader market. Financial terms in such collaborations can often include milestone-based payments, equity stakes, and joint ventures.

Beyond corporate and venture investments, public funding and government grants have emerged as crucial for sustaining long-term research and development in autonomous vehicle technology. Several governments recognize the potential economic and societal benefits of AVs and have initiated various funding programs to promote innovation in this field. These grants are frequently aimed at reducing the financial risk for early-stage companies and fostering a collaborative research environment.

Special Purpose Acquisition Companies (SPACs) have recently become a prevalent funding mechanism for AV startups looking to go public. SPACs offer a streamlined path to the public markets, which can accelerate capital accumulation and allow the company to expand more rapidly. However, the efficacy and long-term sustainability of SPAC-funded ventures are subjects of ongoing debate among financial experts.

Private equity also plays a role, albeit more commonly in the mid-to late-stage funding rounds. Private equity firms look for companies that have moved beyond the initial development phase and demonstrated a stable path to profitability. Their involvement often brings not only capital but also strategic oversight and operational expertise, ensuring the firm is well-positioned for scalable growth.

Institutional investors like pension funds, mutual funds, and hedge funds have been somewhat conservative yet increasingly open to investing in autonomous vehicle technology. Due to the longer timelines for ROI, these institutions generally focus on companies with a proven track record or significant competitive advantage, thereby ensuring more stable investment outcomes.

Crowdfunding has also emerged as an alternative avenue for raising funds, democratizing the investment process by allowing retail investors to participate in the early stages of an exciting technological journey. Platforms like Kickstarter and Indiegogo have hosted successful campaigns for AV-related projects, although these are generally more suitable for smaller-scale innovations or niche products rather than full-fledged autonomous vehicle systems.

Another noteworthy trend is the rise of accelerators and incubators specifically targeting AV startups. Programs like Y Combinator, Techstars, and specialized AV-focused accelerators provide not just funding but also mentorship, resources, and network access. Startups accepted into these programs often leave with not only financial backing but also a much clearer pathway to market entry and scale.

Sovereign wealth funds from oil-rich nations have increasingly found autonomous vehicle technology appealing as part of their strategy to diversify national income sources away from fossil fuels. These funds bring significant capital and are often willing to engage in long-term investments that align with their national economic visions.

However, despite the influx of capital, the investment landscape is not without its challenges. The highly experimental nature of AV technology means that many startups face prolonged periods without revenue. This scenario necessitates a robust financial strategy to sustain operations through research, development, and testing phases that could last years. Moreover, the rapid pace of technological advancement means that these companies must be continuously innovative and adaptable to avoid obsolescence.

Regulatory ambiguities also pose a significant risk to investors. Laws around autonomous driving vary significantly across regions and are still evolving. This uncertainty can affect the scalability of AV technology and its market penetration, making it a critical factor in investment decisions. Companies with a keen understanding of regulatory landscapes and strong lobbying efforts stand a better chance of securing substantial funding.

Intellectual property (IP) considerations are another facet of the investment landscape. AV startups are engaged in a technological arms race, with patents often being the battlegrounds. Strong IP portfolios can be vital in securing investment, as they provide an additional layer of security by protecting the technological innovations from being easily replicated by competitors.

In summation, the investment and funding landscape for startups and innovators in the autonomous vehicle sector is as varied as it is dynamic. Multiple financial avenues, from venture capital and corporate partnerships to public grants and SPACs, support these companies on their journeys. The rapid evolution of AV technology promises significant returns but also demands careful navigation through financial, regulatory, and operational challenges. As the sector continues to grow, so too will the sophistication and variety of its funding mechanisms, reflecting the boundless potential and inherent risks of a future driven by autonomous vehicles.

Chapter 24:
Long-Term Predictions and Trends

As we look ahead, the landscape of autonomous driving is set to experience transformative shifts. We're likely to witness a merging of AI advancements and human-centric design, with vehicles becoming more intelligent and responsive to the needs of their passengers. Electrification will go hand-in-hand with autonomy, driving down emissions and fostering a sustainable transport ecosystem. Societal impacts, including the redesign of urban areas to accommodate self-driving cars and the potential reduction in personal car ownership, will trigger profound changes in how we move. Security concerns, though prevalent, will be addressed with more sophisticated cybersecurity measures. Ultimately, long-term trends will revolve around creating a seamless, efficient, and safe autonomous transport network that reshapes our daily lives.

Technological Advancements

The evolution of autonomous driving technology has been nothing short of remarkable, driven by a confluence of advancements in artificial intelligence, sensor capabilities, and data analytics. These innovations are laying the groundwork for a future where not just cars, but entire transportation systems, operate with unprecedented levels of autonomy and efficiency. At the core of these transformations are sophisticated AI algorithms that can learn from vast amounts of data, enabling vehicles to perceive and navigate complex environments.

AI at the Wheel

One of the most pivotal technological advancements is in the realm of machine learning and AI. Initially, self-driving cars relied heavily on rule-based algorithms that mandated explicit instructions for every conceivable scenario. However, the modern approach leverages deep learning, enabling vehicles to process and interpret a massive array of sensory inputs, such as radar, LIDAR, and camera feeds, in real-time. This evolution allows for an adaptation to an almost infinite number of driving conditions, elevating the reliability and safety of autonomous systems.

Another critical advancement is in sensor technology. The suite of sensors integrated into autonomous vehicles, including LIDAR, radar, and ultrasonic sensors, continually evolves. Enhancements in these technologies have significantly improved object detection and ranging capabilities, which are crucial for safe navigation. LIDAR, for example, uses light to measure distances to objects, creating precise 3D maps of the vehicle's surroundings. Ongoing research aims to make these sensors more cost-effective and compact, which will be essential for mass deployment.

Connectivity forms yet another pillar of technological advancement. Vehicle-to-everything (V2X) communication systems are being developed to enable cars to interact with each other and the infrastructure around them. Such systems can share information about traffic conditions, road hazards, and optimal routes, making collective navigation safer and more efficient. This interconnectedness transforms isolated autonomous vehicles into part of an intelligent, integrated transport ecosystem.

Powering these technological advancements is the improved computational power available in modern vehicles. High-performance computing units capable of executing complex AI algorithms in real-time are now small enough to be fitted into cars. These processors can handle immense datasets and perform intricate computations that were previously unthinkable in a mobile environment. The development of custom AI chips also promises to improve energy efficiency, an important consideration as electric vehicles become more prevalent.

Cloud computing and big data analytics also play an instrumental role. Fleets of autonomous vehicles generate massive amounts of data, covering virtually every aspect of their operation. This data is invaluable for refining algorithms and improving system performance. Advanced analytics and real-time processing enable quicker iterations and deployments of improved models. Moreover, cloud platforms allow for the rapid dissemination of updates and insights across entire networks of vehicles, ensuring consistent and up-to-date performance.

One of the more nuanced advancements is in the realm of human-machine interaction (HMI). Ensuring that users are comfortable and can easily interact with autonomous systems is crucial for mass adoption. Innovations in user interface design aim to make these interactions as intuitive and seamless as possible. From voice-controlled commands to advanced driver monitoring systems, efforts are underway to ensure that the transition from manual to autonomous driving is smooth and user-friendly.

Additionally, advancements in simulation and virtual testing environments cannot be overlooked. These platforms allow for the exhaustive testing of autonomous systems under a variety of conditions without the risks associated with real-world testing. Simulations can replicate rare and dangerous scenarios, providing vital data that helps in refining autonomous driving systems to handle extreme cases with greater competence.

In the domain of cybersecurity, significant strides are being made to safeguard autonomous vehicles against a myriad of threats. As vehicle connectivity increases, so does the risk of cyber-attacks. Advanced encryption techniques and secure communication protocols are being developed to protect the integrity and confidentiality of data exchanged between vehicles and infrastructure. Continuous monitoring and quick response mechanisms are also crucial to mitigate any potential vulnerabilities in real-time.

Battery technology and energy management systems have also seen noteworthy improvements. The move towards autonomy often goes hand-in-hand with electric vehicle technology. Innovations in battery materials, charging infrastructure, and energy-efficient components are crucial for extending the range and reducing the downtime of autonomous electric vehicles. Better energy management not only enhances vehicle performance but also aligns with broader sustainability goals.

The realm of software development for self-driving cars has become a field in itself, populated with specialized tools and platforms designed to streamline the creation, testing, and deployment of autonomous systems. Open-source platforms, standardization of protocols, and collaborative development efforts are accelerating progress by allowing developers to build on each other's work rather than starting from scratch.

Robust frameworks for the continuous monitoring and validation of AI systems in autonomous vehicles are being established to ensure safety and reliability. These frameworks include rigorous validation protocols, performance metrics, and safety assessments that autonomous vehicles must pass before they are deemed roadworthy. This comprehensive approach to validation helps build public trust and accelerates regulatory approval processes.

In conclusion, the technological advancements driving the future of autonomous vehicles are vast and multi-faceted. They encompass improvements in AI and machine learning, sensor technology, connectivity, computational power, cloud computing, human-machine interaction, simulation, cybersecurity, battery technology, software development, and validation frameworks. Each of these areas is essential for creating a reliable, efficient, and safe autonomous transportation ecosystem, pushing us closer to a future where autonomy in driving is not just a concept, but a seamless, everyday reality.

Societal Changes

The advent of autonomous driving will significantly alter the societal landscape, touching aspects of our daily lives and broader social structures. One of the most immediate societal changes will be seen in the dynamics of personal transportation. With autonomous vehicles (AVs) becoming more prevalent, traditional concepts of car ownership may evolve. The appeal of owning a car may diminish as more people opt for on-demand, ride-hailing services operated by fleets of autonomous vehicles. This shift could lead to a marked decrease in the number of privately owned vehicles, impacting everything from car manufacturing to urban planning.

Consider the implications for urban areas. Currently dominated by car parks and traffic congestion, cities might undergo radical transformations. Parking lots could become obsolete, freeing up valuable real estate for other uses such as parks, bike lanes, and expanded pedestrian zones. This can contribute to more liveable, greener, and aesthetically pleasing urban environments. Additionally, as congestion decreases, the efficiency of travel within cities will improve, making daily commutes shorter and less stressful.

Societal changes will also manifest more subtly in people's everyday lives. For instance, families without the traditional family car might save money that they could allocate to other priorities like education or leisure activities. This could, in turn, boost other sectors of the economy. Furthermore, the daily commute will become a time for productivity or relaxation, fundamentally altering how people schedule their day. Imagine reading, working, or even sleeping during what used to be an hour-long, stress-inducing commute.

Another significant change will come in the form of accessibility. For the elderly and those with disabilities, autonomous vehicles promise newfound independence. These populations, often limited by conventional transportation, will now have greater freedom to move about, thereby enriching their social and professional lives. The implications extend beyond mere convenience; increased mobility can improve mental health, social engagement, and overall quality of life for these communities.

Interestingly, the move to autonomous vehicles could also impact social interactions. While sharing rides with strangers isn't a new concept, the nature of these interactions could change. In autonomous shuttles or ride-sharing services, the absence of a human driver might make the overall experience less personal but more efficient. Dependable, predictable service without the variability introduced by human drivers could appeal to many users.

However, it's not all positive. Autonomous driving technology brings its array of societal dilemmas and ethical questions. Public trust in technology remains a potent issue. Surveys consistently show a mixed reception to autonomous driving technologies, often stemming from fears about safety, cybersecurity, and the potential for job losses in industries such as trucking and taxi services. As society wrestles with these changes, policymakers will have their work cut out for them in balancing innovation with public sentiment.

When we talk about societal change, we can't ignore the potential ramifications for social equity. Access to autonomous vehicles or ride-sharing services might initially be limited to affluent areas or high-income individuals, thereby exacerbating existing socioeconomic divides. Policymakers and companies will need to work together to ensure these services are accessible to a broad range of people, including those in underserved communities.

Moreover, let's delve into the workplace. Jobs that currently exist for millions of drivers, from the logistics sector to ride-sharing, could diminish or transform drastically. While new jobs will emerge in technology, maintenance, and fleet management, they will require different skill sets. The transition phase could lead to social unrest if not managed carefully. Governments and educational institutions will need to invest in reskilling programs to prepare the workforce for this change. Economic disparities could worsen if these measures are not implemented effectively.

In the broader societal context, the concept of time and space will transform. Long-distance travel might become more seamless as autonomous vehicles reduce the strain of driving. This could make rural or suburban living more appealing, as the daily grind of commuting becomes less taxing. Urban sprawl could take on new dimensions as people choose to live farther from city centers without the penalty of a long, tedious commute.

Furthermore, the ripple effects of these social shifts could permeate through various other sectors. Real estate markets might adapt to new transportation realities, potentially affecting housing prices and regional development patterns. Communities built around car-centric lifestyles will need to reimagine their infrastructure to accommodate autonomous vehicles.

Even the educational landscape might see shifts as transportation becomes more accessible and efficient. Students could commute from greater distances, opening up educational opportunities that were previously impractical due to transportation barriers. This could foster greater educational equity and diversification of talent pools within academic institutions.

The interplay between autonomous driving technology and societal norms will also extend into family dynamics. Parents, no longer required to chauffeur their children to school or extracurricular activities, might have more time for other responsibilities or leisure activities. This newfound availability of time could shift the balance of domestic responsibilities, potentially offering a more balanced family dynamic.

Moreover, cities and towns might become less vehicle-centric over time. Initiatives aimed at reducing congestion and better utilizing space will differ based on regional needs. In North America, large suburban areas might see more shared autonomous vehicle services, while European cities might prioritize integrating autonomous vehicles with existing public transit systems. These diverse approaches will shape the unique societal outcomes experienced in different regions.

Transparency and public engagement will be crucial as these changes unfold. Governments and companies will need to work hand-in-hand to create rules and guidelines that not only foster technological advancements but also consider social impact. Securing public trust will be an ongoing process requiring open communication, consistent safety protocols, and clear regulations.

Lastly, let's not overlook the cultural aspects. Autonomous driving will very likely influence elements of popular culture, from movies and novels to art and social media. Society often mirrors and anticipates technological trends in its cultural output. Expect the rise of a new genre of fiction that explores the implications of a driver-less world, as well as real-world, grassroots movements that either support or oppose the proliferation of autonomous vehicles.

In conclusion, the societal changes ushered in by the advent of autonomous driving will be multifaceted and profound. From altering urban landscapes to impacting everyday routines, the effects will be widespread. Navigating these changes will require collective effort, thoughtful policymaking, and a willingness to adapt. While the road ahead may be complex, the potential benefits for society are vast, making it a transformative journey worth embarking upon.

Chapter 25:
How to Prepare for an Autonomous Future

As we edge closer to an era dominated by autonomous vehicles, it's crucial to adapt both our skills and strategic approaches to seize the opportunities this transformative technology presents. Education systems need to embrace curricula reflecting AI, robotics, and data analytics to equip the future workforce. Businesses must pivot, integrating cutting-edge technologies and reshaping their models to remain competitive in an AI-driven market. Additionally, policymaking and regulatory frameworks will require agile adaptation to address the nuanced challenges emerging from autonomous systems. By staying informed and proactive, individuals and organizations can navigate the complexities of an autonomous future and harness its full potential.

Skills and Education

Preparing for an autonomous future requires more than just technological advancements; it demands a significant shift in skills and education. With autonomous vehicles becoming increasingly integral to the transportation landscape, a well-prepared workforce is essential for both their development and maintenance. This preparation spans various sectors, encompassing engineering, data science, cybersecurity, policy-making, and even ethics.

First and foremost, the field of engineering needs to adapt to the requirements of autonomous technologies. Mechanical engineers, traditionally focused on combustion engines and mechanical components, must now grasp the intricacies of software-driven systems and electronic sensors. Courses that once focused on thermodynamics and material science should now incorporate elements of artificial intelligence, machine learning algorithms, and sensor fusion techniques.

Similarly, electrical and computer engineers must upgrade their competencies. Understanding the interplay between hardware and software is no longer optional. As these vehicles rely heavily on advanced electronics and cloud computing, there is a growing need for expertise in areas like embedded systems, real-time operating systems, and network communications. Educational institutions must pivot accordingly, offering specialized courses and hands-on labs that emphasize these new technologies.

Data science and AI are at the core of autonomous driving technology. Professionals in these fields must not only master machine learning algorithms but also understand their applications in real-world scenarios. For instance, they should be well-versed in training neural networks to recognize traffic signs, pedestrians, and other vehicles. Big data plays a significant role here, and those aspiring to work in this field should be proficient in data analytics, cloud computing, and real-time data processing.

Cybersecurity is another critical area requiring immediate attention. Autonomous vehicles are essentially computers on wheels, susceptible to various cyber threats. Protecting these systems from hacking and ensuring secure communication between vehicles and infrastructure is paramount. Educational programs must place a strong emphasis on security protocols, encryption techniques, and ethical hacking practices.

Policy-making and regulatory affairs can't be overlooked. As governments and organizations draft new regulations to keep pace with technological advances, there is a pressing need for professionals who understand both the technical and legal landscapes. Courses in public policy, law, and ethics should cover topics specific to autonomous driving, including liability in case of accidents, data privacy concerns, and the ethical implications of AI decision-making.

The ethical dimension of autonomous vehicles is crucial. Engineers and policymakers alike must grapple with questions of moral responsibility. When an autonomous vehicle faces an unavoidable accident, who is to blame—the manufacturer, the software developer, or the owner? Integrating ethics into the curriculum can help aspiring professionals navigate these challenging moral landscapes.

Moreover, interdisciplinary training is increasingly important. While specialized knowledge is essential, the ability to collaborate with professionals from different backgrounds can lead to more innovative and comprehensive solutions. For example, urban planners working alongside transportation engineers and data scientists can design smarter cities equipped to handle autonomous vehicles efficiently.

Upskilling the current workforce is just as critical as educating the new generation. For professionals already in the field, continuous learning opportunities such as certifications, online courses, and workshops can provide the necessary skills to stay current. Many institutions now offer modular courses that can be completed part-time, enabling professionals to learn without sacrificing their existing commitments.

As for the broader public, educational initiatives should not be confined to technical realms alone. Understanding the basics of autonomous technology can help in its societal acceptance. Public seminars, online courses, and outreach programs can demystify these technologies, helping people feel more comfortable and informed about the future.

Schools and universities must lay the groundwork for this transformation as early as possible. Introducing coding and basic robotic principles at the K-12 level can spark interest in younger students. Competitions, hands-on projects, and interactive learning modules can make these subjects more engaging, preparing the next generation for a career in autonomous technology.

The role of educators is also evolving. Teachers and professors must themselves stay up-to-date with the latest technological advances. This requires institutions to invest in faculty training and development. Collaborative partnerships with industries can provide educators with firsthand experience and insights, which they can then pass on to their students.

Internships and cooperative education programs will play a vital role in preparing students for real-world challenges. Collaborating with companies pioneering in autonomous technologies can provide valuable practical experience. These programs can help bridge the gap between theoretical knowledge and practical application, making graduates more attractive to potential employers.

Finally, fostering a culture of lifelong learning is essential. The pace of technological change shows no signs of slowing, and today's cutting-edge knowledge may become obsolete tomorrow. Encouraging a mindset of continuous improvement and adaptability can prepare professionals to navigate an ever-evolving landscape.

In summary, equipping ourselves with the right skills and knowledge is fundamental to preparing for an autonomous future. Whether it's through formal education, professional development, or public outreach, a comprehensive approach is necessary to ensure we are ready for what lies ahead. This preparation will not only smooth the transition to autonomous vehicles but also harness their full potential for societal benefit.

Business Strategies

The coming wave of autonomous vehicles is more than just a technological revolution — it represents a seismic shift in business strategies across industries. Companies must adapt swiftly or risk obsolescence. Key players in various sectors are already rethinking their approaches to remain competitive and relevant in a future dominated by self-driving cars.

One of the initial steps businesses need to take is investing in AI and machine learning research. It's not just about having the technology but understanding and leveraging it to create new value propositions. Companies like Tesla and Waymo have shown that innovation in autonomous driving technology can drive significant market capitalizations. However, it's not enough to deploy advanced technologies; enterprises must also establish robust frameworks for ongoing development and improvement.

Integrating autonomous vehicles into a business model requires wrestling with considerable infrastructural changes. Firms need to develop or upgrade their digital infrastructure to handle the complex software demands and high data throughput that self-driving cars necessitate. This involves transitioning from legacy systems to more scalable and modular software architectures. Moreover, cloud computing services and data storage solutions become pivotal in this transformation, enabling real-time processing and analytics.

Partnerships and collaborations will also be crucial moving forward. Car manufacturers, technology firms, insurance companies, and even city planners must work together to create a cohesive ecosystem. These alliances can facilitate shared standards and protocols, making it easier to scale autonomous solutions. Such collaborative efforts can also spread the considerable costs associated with autonomous vehicle deployment.

Companies must also pivot their marketing and customer engagement strategies. Understanding public sentiment, educating potential clients about the benefits and safety of autonomous vehicles, and addressing their concerns become essential. Businesses will need to engage in public relations campaigns, social media outreach, and perhaps even interactive demonstrations to win trust and market share.

For traditional automotive companies, there's a pressing need to diversify. While vehicle sales remain important, services related to autonomous driving will likely become significant revenue streams. Subscription models for driverless car services, remote vehicle management, and AI-driven maintenance programs could all contribute to future profitability. Diversification can help mitigate risks and tap into new, emerging markets.

Financial strategies will need recalibrating as well. High upfront development costs can discourage direct investment in autonomous technology. However, venture capital has shown a strong interest in backing innovative startups in this sector. Businesses should be prepared to leverage venture capital funding, public-private partnerships, and even government grants to fuel their autonomous vehicle initiatives.

Legal and regulatory compliance is another critical factor. Each region may have its own rules governing the deployment and operation of autonomous vehicles. Companies must allocate resources to navigate these regulatory landscapes and advocate for favorable legislative frameworks. This will ensure smoother market entry and operational sustainability.

Human capital is also pivotal. Upskilling and educating the current workforce to handle the new demands posed by autonomous technology can't be overlooked. Offering training programs on AI, machine learning, cybersecurity, and software development is paramount. Companies should foster an environment that encourages continuous learning and innovation. Additionally, acquiring talent well-versed in these technologies can provide a competitive edge.

Supply chain considerations will also come into play. Autonomous vehicles will likely require new types of sensors, advanced semiconductor chips, and specialized components. Thus, reevaluating existing supply chain relationships and forming new ones is essential for smooth production cycles. Organizations should plan logistic strategies that accommodate the unique needs of autonomous vehicle manufacturing and maintenance.

The ride-sharing industry is poised for a dramatic transformation with the advent of autonomous vehicles. Market leaders like Uber and Lyft are already investing heavily in self-driving technology, recognizing that the future of on-demand mobility is driverless. These companies should focus on optimizing their fleet management algorithms, creating sophisticated pricing models, and ensuring the safety and reliability of their autonomous services.

Retail and logistics sectors are equally looking at revolutionary changes. Businesses can integrate self-driving delivery vehicles into their logistics networks to enhance efficiency and reduce costs. For instance, Amazon has been exploring autonomous delivery vans and drones, signaling the enormous potential for cost savings and improved delivery times. Retailers should prepare for a future where autonomous vehicles handle most last-mile deliveries.

As urban landscapes evolve, real estate and commercial property sectors must adapt to the spatial implications of autonomous vehicles. Self-driving cars will change parking needs and traffic flows, potentially opening up urban spaces for new commercial developments. Real estate companies should consider how these changes will affect property values and leasing arrangements. They may also need to invest in building infrastructure that supports autonomous driving technologies, like dedicated drop-off zones and charging stations.

Emerging business models in autonomous transportation will require an agile approach to product management and development. Agile methodologies, particularly iterative development cycles, could prove invaluable. This approach allows companies to quickly adapt to technological advancements and shifts in consumer behavior, maintaining a competitive edge in a rapidly evolving market.

Finally, companies should not disregard the enormous influence of big data in shaping business strategy. Autonomous vehicles generate massive amounts of data that can provide significant business insights. Real-time data analytics can improve decision-making processes across the board, from optimizing routes and enhancing safety features to tailoring personalized experiences for users. Investing in robust data analytics platforms and hiring skilled data scientists should be part of any forward-looking business strategy in this arena.

Ultimately, preparing for an autonomous future involves a multifaceted strategy that touches every aspect of business operations. From technological investments and infrastructural upgrades to regulatory compliance and workforce training, companies must be comprehensive in their approach. The road ahead is challenging but filled with opportunities for those willing to innovate and adapt.

Conclusion

The journey through the world of autonomous driving has been both fascinating and enlightening. With each chapter in this book, we've explored the multifaceted dimensions of a transportation revolution that's not just on the horizon but already reshaping our roads, cities, and lives. Now, as we stand at the intersection of technology and humanity, it's crucial to reflect on the overarching themes and future implications of autonomous vehicles.

First and foremost, the technological marvels driving self-driving cars—sensors, AI algorithms, and V2X communication—are nothing short of revolutionary. These advancements have made what once seemed like science fiction a tangible reality, and they continue to evolve at a breakneck pace. However, it's not just the technology itself but the synergy between various innovations that holds the promise of safer, more efficient, and environmentally friendly transportation.

On a societal level, the implications are profound. Autonomous vehicles are poised to redefine urban landscapes, potentially reducing congestion and changing the dynamics of public and private transportation. They promise to improve accessibility for those who are unable to drive due to age or disability, thereby promoting greater inclusivity. Yet, these changes also bring challenges—public perception and social acceptance are critical factors that will influence the rate and extent of autonomous vehicle adoption.

The economic landscape will also witness significant shifts. Jobs in transportation and related sectors may be displaced, but new opportunities in tech, data analytics, and vehicle maintenance will emerge. Market growth in autonomous vehicles is expected to be robust, offering lucrative opportunities for early movers and innovators. Public policy and regulation will play a crucial role in shaping this market, ensuring it evolves in a manner that is both inclusive and equitable.

Ethics and legality are perhaps the most contentious aspects of the autonomous driving debate. Moral dilemmas, such as decision-making in crash scenarios, pose difficult questions with no easy answers. Similarly, the legal frameworks governing responsibility and liability in accidents will need to be continually updated to keep pace with technological advancements. The role of Big Data in autonomous driving also raises significant privacy and security concerns that must be addressed to build public trust.

From a safety perspective, autonomous vehicles hold the promise of reducing the number of road accidents, many of which are caused by human error. However, this potential can only be realized through rigorous validation, testing protocols, and advancements in collision avoidance systems. The interplay between human drivers and autonomous systems will also be crucial in this transition phase.

Environmentally, self-driving cars offer the potential to significantly reduce emissions through optimized driving patterns and the use of sustainable materials. Yet, the extent of this impact depends on the broader adoption of electric vehicles and the development of cleaner energy sources.

Looking ahead, the future of public and personal transportation seems poised for a radical transformation. Autonomous buses, shuttles, and ride-sharing services may become the norm, effectively blending into existing public transportation systems. This evolution will depend heavily on advances in fleet management, navigation systems, and human-machine interaction.

Startups and innovators will continue to play a pivotal role in this landscape, driving disruptive technologies and attracting significant investment. Collaboration between tech companies, automotive manufacturers, and policymakers will be essential to navigate the complexities of this rapidly evolving field.

Education and skill development will be crucial for preparing the workforce for an autonomous future. As business strategies evolve to incorporate these new technologies, companies will need to be agile and forward-thinking. For individuals, understanding the basics of AI and autonomous systems will be increasingly important, not just for career opportunities but for informed citizenship.

Ultimately, the promise of autonomous driving lies in its potential to make transportation safer, more efficient, and environmentally sustainable. However, realizing this potential will require a concerted effort from all stakeholders—technologists, policymakers, industry professionals, and the public. Only through such collaborative effort can we navigate the ethical, legal, and societal challenges that lie ahead.

As we've explored in this book, the journey to achieving widespread autonomous driving is complex and multifaceted. It's a journey that requires balancing innovation with responsibility, optimism with pragmatism, and technological advancement with human values. As we continue to push the boundaries of what's possible, we must remain vigilant in ensuring that this new era of transportation serves the greater good.

In conclusion, the future of autonomous driving is not just about cars and technology; it's about the society we aspire to create. A society where mobility is a right, not a privilege, and where technology serves humanity in the most profound ways. The road ahead may be challenging, but it is also filled with immense possibilities. Together, we can build a future where the promise of autonomous vehicles is fully realized, transforming our world for the better.

Appendix A:
Appendix

The appendix serves as an essential component of the book, offering readers a treasure trove of supplementary materials, detailed explanations, and additional resources that augment the main text. Here, you'll find comprehensive tables that break down complex data, charts that illustrate trends over time, and extended case studies that didn't quite fit into the main chapters but offer invaluable insights. Whether you're looking for a deeper dive into sensor technologies, historical timelines, or ethical frameworks discussed briefly in the book, this appendix provides the additional context, data, and references necessary to fully grasp the transformative power of autonomous driving. By curating a well-rounded collection of technical details, legislative documents, research papers, and further readings, this section ensures that both casual readers and industry professionals can enhance their understanding and stay informed on the cutting-edge developments in AI and transportation.

Glossary of Terms

This glossary provides definitions and explanations of key terms used throughout the book, aiming to offer clarity on concepts related to autonomous driving and AI in transportation. Understanding these terms will enhance your comprehension of the material presented in the chapters.

A

AI (Artificial Intelligence) - The simulation of human intelligence processes by machines, especially computer systems.

Autonomous Vehicle (AV) - A self-driving car that operates without human intervention through the use of sensors, algorithms, and AI.

C

Collision Avoidance System - A safety feature that detects potential collisions and takes action to prevent accidents.

Connectivity - The ability of a vehicle to communicate with other systems or devices, often through the internet.

D

Deep Learning - A subset of machine learning that utilizes complex neural networks to process data at multiple levels of abstraction.

Driver Assist - Technologies that assist human drivers in the driving process but don't take over full control.

E

Edge Computing - Processing data near the source of data generation rather than in a centralized data center, crucial for real-time applications like autonomous driving.

F

Fleet Management - The management of a fleet of vehicles to improve efficiency, monitor performance, and ensure compliance with regulations.

I

Infrastructure - The physical and organizational structures needed for the operation of autonomous vehicles, including roads, communication networks, and charging stations.

Internet of Things (IoT) - The network of interconnected devices that communicate and share data with each other, enhancing vehicle functionality.

L

LIDAR (Light Detection and Ranging) - A sensor technology that measures distance by illuminating a target with a laser and analyzing the reflected light.

Localization - The process of determining a vehicle's exact location using data from various sensors and maps.

M

Machine Learning - A branch of AI that enables systems to learn from data and improve their performance over time without being explicitly programmed.

N

Neural Network - A series of algorithms that attempt to recognize patterns in data, mirroring the way a human brain operates.

P

Predictive Maintenance - Techniques that use data analysis tools to predict when vehicle components will fail and when maintenance should be performed.

R

Regulation - Laws and rules governing the development, testing, and deployment of autonomous vehicles.

S

Sensors - Devices that detect events or changes in the environment and send this information to other electronics, often used for navigation and safety in AVs.

Smart Grid - An electricity supply network that uses digital communications technology to detect and react to local changes in usage, helping to enable more efficient vehicle charging.

V

Vehicle-to-Everything (V2X) Communication - A system where vehicles communicate with each other and with infrastructure to enhance safety and traffic efficiency.

This glossary covers primary terms essential for grasping the discussions on autonomous driving and AI. As you continue through the book, refer back to these definitions to reinforce your understanding of the complex and evolving landscape of autonomous transportation.

Further Reading and Resources

As we've navigated through the comprehensive glossary of terms related to autonomous driving, it becomes clear that there is always more to learn. The landscape of AI, machine learning, and self-driving technology is constantly evolving, and remaining informed can be a challenge. However, there are numerous resources available to deepen your understanding and keep abreast of the latest developments in this dynamic field.

Books offer an extensive examination of topics related to autonomous vehicles. Titles such as "Artificial Intelligence: A Guide for Thinking Humans" by Melanie Mitchell and "The Age of Em: Work, Love, and Life When Robots Rule the Earth" by Robin Hanson, provide deep insights into the broader AI landscape and its implications. For those specifically interested in autonomous vehicles, "Autonomy: The Quest to Build the Driverless Car—and How It Will Reshape Our World" by Lawrence D. Burns is a must-read that delves into the history, technology, and future of driverless cars.

Journal articles and research papers are indispensable for those wanting to dive deep into academic and technical discussions. Publications such as IEEE Intelligent Transportation Systems Magazine and the Journal of Field Robotics regularly feature cutting-edge research on AI, sensors, machine learning algorithms, and safety protocols in autonomous driving. These periodicals are essential for staying current with the latest scientific advancements and understanding the technical challenges that researchers and engineers are working to overcome.

Industry reports and whitepapers from organizations like McKinsey & Company, Gartner, and the RAND Corporation provide detailed analyses and forecasts on market trends, economic impacts, and societal implications of autonomous driving technologies. These documents are particularly useful for business professionals and policymakers, offering a well-rounded view of how self-driving cars might transform various sectors.

The rise of digital media has also spawned a plethora of online courses and webinars that cater to both beginners and advanced learners. Platforms like Coursera, edX, and Udacity offer specialized courses on autonomous driving, machine learning, and AI ethics. These courses, often created in collaboration with top universities and tech companies, provide valuable hands-on experience and insights from industry experts.

For a more informal learning experience, podcasts and YouTube channels serve as accessible resources. Podcasts such as "The AI Alignment Podcast" and "Autonocast" feature interviews with leading experts and discussions on the latest trends and challenges in the autonomous vehicle space. Similarly, YouTube channels like "Two Minute Papers" and "Computerphile" break down complex topics in a digestible format, making them ideal for quick yet informative learning sessions.

Professional networks and conferences also play a crucial role in advancing knowledge and fostering collaboration. Attend events such as the International Conference on Intelligent Transportation Systems (ITS) or the Consumer Electronics Show (CES) to engage with pioneers in the field and witness the unveiling of the latest innovations. Membership in organizations like the Institute of Electrical and Electronics Engineers (IEEE) or the Association for the Advancement of Artificial Intelligence (AAAI) provides access to exclusive resources, publications, and networking opportunities.

Government websites and regulatory bodies often publish guidelines, standards, and research findings that are invaluable for understanding the legal and policy framework surrounding autonomous vehicles. The U.S. Department of Transportation (DOT), the National Highway Traffic Safety Administration (NHTSA), and their international counterparts regularly update their websites with reports, legislative texts, and safety assessments pertinent to self-driving technology.

Moreover, active participation in online forums and discussion groups can be immensely beneficial. Platforms like Reddit, Stack Exchange, and specialized LinkedIn groups host vibrant communities of experts and enthusiasts exchanging ideas, solving problems, and sharing the latest news. Engaging in these discussions can offer real-time insights and practical advice that might not be readily available in more formal resources.

Lastly, it's worth keeping an eye on the official blogs and newsletters of leading tech companies and startups in the autonomous driving space. Companies such as Waymo, Tesla, and NVIDIA frequently publish updates on their research, technological advancements, and industry perspectives. Subscribing to these communications ensures that you receive firsthand information directly from pioneers pushing the boundaries of what's possible with AI and autonomous vehicles.

To sum up, the realm of autonomous driving is vast and ever-changing. Tapping into a diverse set of resources—from academic journals and industry reports to online courses and community forums—can help anyone interested in this transformative technology stay informed and engaged. Whether you are a researcher, industry professional, policymaker, or simply an enthusiast, there are abundant avenues for advancing your knowledge and contributing to the ongoing dialogue about the future of transportation. By continuing to explore and utilize these resources, we can collectively shape a safer, more efficient, and innovative era of autonomous vehicles.

www.ingramcontent.com/pod-product-compliance
Lightning Source LLC
Chambersburg PA
CBHW051233050326
40689CB00007B/904